P9-AOB-262

Audel™

Automated Machines and Toolmaking
All New 5th Edition

WILDER BRANCH LIBRARY
7140 E. SEVEN MILE RD.
DETROIT, MI 48234

Audel™

Automated Machines and Toolmaking
All New 5th Edition

Rex Miller
Mark Richard Miller

Wiley Publishing, Inc.

Vice President and Executive Group Publisher: Richard Swadley
Vice President and Executive Publisher: Robert Ipsen
Vice President and Publisher: Joseph B. Wikert
Executive Editorial Director: Mary Bednarek
Editorial Manager: Kathryn A. Malm
Executive Editor: Carol A. Long
Senior Production Manager: Fred Bernardi
Development Editor: Kevin Shafer
Production Editor: Vincent Kunkemueller
Text Design & Composition: TechBooks

Copyright © 2004 by Wiley Publishing, Inc. All rights reserved.
Copyright © 1965, 1970, and 1978 by Howard W. Sams & Co., Inc.
Copyright © 1983 by The Bobbs-Merrill Co., Inc.
Copyright © 1986 by Macmillan Publishing Company, a division of Macmillan Inc.

Published simultaneously in Canada

No part of this publication may be reproduced, stored in a retrieval system, or transmitted in any form or by any means, electronic, mechanical, photocopying, recording, scanning, or otherwise, except as permitted under Section 107 or 108 of the 1976 United States Copyright Act, without either the prior written permission of the Publisher, or authorization through payment of the appropriate per-copy fee to the Copyright Clearance Center, Inc., 222 Rosewood Drive, Danvers, MA 01923, (978) 750-8400, fax (978) 646-8600. Requests to the Publisher for permission should be addressed to the Legal Department, Wiley Publishing, Inc., 10475 Crosspoint Blvd., Indianapolis, IN 46256, (317) 572-3447, fax (317) 572-4447, E-mail: permcoordinator@wiley.com.

Limit of Liability/Disclaimer of Warranty: While the publisher and author have used their best efforts in preparing this book, they make no representations or warranties with respect to the accuracy or completeness of the contents of this book and specifically disclaim any implied warranties of merchantability or fitness for a particular purpose. No warranty may be created or extended by sales representatives or written sales materials. The advice and strategies contained herein may not be suitable for your situation. You should consult with a professional where appropriate. Neither the publisher nor author shall be liable for any loss of profit or any other commercial damages, including but not limited to special, incidental, consequential, or other damages.

For general information on our other products and services, please contact our Customer Care Department within the United States at (800) 762-2974, outside the United States at (317) 572-3993, or fax (317) 572-4002.

Trademarks: Wiley, the Wiley Publishing logo, Audel, and related trade dress are trademarks or registered trademarks of John Wiley & Sons, Inc., and/or its affiliates in the United States and other countries, and may not be used without written permission. All other trademarks are the property of their respective owners. Wiley Publishing, Inc., is not associated with any product or vendor mentioned in this book.

Wiley also publishes its books in a variety of electronic formats. Some content that appears in print may not be available in electronic books.

Library of Congress Cataloging-in-Publication Data:

ISBN: 0-764-55528-6

Printed in the United States of America

10   9   8   7   6   5   4   3   2   1

Contents

Acknowledgments

A number of companies have been responsible for furnishing illustrative materials and procedures used in this book. At this time, the authors and publisher would like to thank them for their contributions. Some of the drawings and photographs have been furnished by the authors. Any illustration furnished by a company is duly noted in the caption.

The authors would like to thank everyone involved for his or her contributions. Some of the firms that supplied technical information and illustrations are listed below:

A. F. Holden Co.

Brown and Sharp Manufacturing Co.

Cincinnati Milacron Co.

Cleveland Automatic Machine Co.

DoAll Co.

Ex-Cell-O Corporation

Federal Products Corp.

Friden, Inc.

Gisholt Machine Co.

Heald Machine Co.

Illinois Gear

Johnson Gas Appliance Co.

L.S. Starrett Co.

Lepel Corporation

Machinery's Handbook, The Industrial Press

Moog Hydro-Point

NASA

Norton Co.

Paul and Beekman Inc.

Sheldon Machine Co.

Thermolyne Corp.

About the Authors

Rex Miller was a Professor of Industrial Technology at The State University of New York—College at Buffalo for over 35 years. He has taught on the technical school, high school, and college level for well over 40 years. He is the author or coauthor of over 100 textbooks ranging from electronics through carpentry and sheet metal work. He has contributed more than 50 magazine articles over the years to technical publications. He is also the author of seven Civil War regimental histories.

Mark Richard Miller finished his B.S. degree in New York and moved on to Ball State University where he obtained the master's and went to work in San Antonio. He taught in high school and went to graduate school in College Station, Texas, finishing the doctorate. He took a position at Texas A&M University in Kingsville, Texas, where he now teaches in the Industrial Technology Department as a Professor and Department Chairman. He has coauthored seven books and contributed many articles to technical magazines. His hobbies include refinishing a 1970 Plymouth Super Bird and a 1971 Roadrunner. He is also interested in playing guitar, which he did while in college as lead in The Rude Boys band.

Introduction

The purpose of this book is to provide a better understanding of the fundamental principles of working with metals in many forms, but with emphasis upon the machining—utilizing both manually operated and automated machines. It is the beginner and the advanced machinist alike who may be able to profit from studying the procedures and materials shown in these pages.

One of the chief objectives has been to make the book clear and understandable to both students and workers. The illustrations and photographs have been selected to present the how-to-do-it phase of many of the machine shop operations. The material presented here should be helpful to the machine shop instructor, as well as to the individual student or worker who desires to improve himself or herself in this trade.

The proper use of machines and the safety rules for using them have been stressed throughout the book. Basic principles of setting the cutting tools and cutters are dealt with thoroughly, and recommended methods of mounting the work in the machines are profusely illustrated. The role of numerically controlled machines is covered in detail with emphasis upon the various types of machine shop operations that can be performed by them.

Some of the latest tools and processes are included. New chapters have been added with updated information and illustrations whenever appropriate. This book, in it's all new fifth edition, has been reorganized into more logical units that can be digested much more easily.

This book has been developed to aid you in taking advantage of the trend toward vocational training of young adults. An individual who is ambitious enough to want to perfect himself or herself in the machinist trade will find the material presented in an easy-to-understand manner, whether studying alone, or as an apprentice working under close supervision on the job.

Chapter 1

Jigs and Fixtures

Jigs and fixtures are devices used to facilitate production work, making interchangeable pieces of work possible at a savings in cost of production. Both terms are frequently used incorrectly in shops. A *jig* is a guiding device and a *fixture* a holding device.

Jigs and fixtures are used to locate and hold the work that is to be machined. These devices are provided with attachments for guiding, setting, and supporting the tools in such a manner that all the workpieces produced in a given jig or fixture will be exactly alike in every way.

The employment of unskilled labor is possible when jigs and fixtures can be used in production work. The repetitive layout and setup (which are time-consuming activities and require considerable skill) are eliminated. Also, the use of these devices can result in such a degree of accuracy that workpieces can be assembled with a minimum amount of fitting.

A jig or fixture can be designed for a particular job. The form to be used depends on the shape and requirement of the workpiece to be machined.

Jigs

The two types of jigs that are in general use are (1) clamp jig and (2) box jig. A few fundamental forms of jigs will be shown to illustrate the design and application of jigs. Various names are applied to jigs (such as drilling, reaming, and tapping) according to the operation to be performed.

Clamp Jig

This device derives its name from the fact that it usually resembles some form of clamp. It is adapted for use on workpieces on which the axes of all the holes that are to be drilled are parallel.

Clamp jigs are sometimes called *open jigs*. A simple example of a clamp jig is a design for drilling holes that are all the same size—for example, the stud holes in a cylinder head (Figure 1-1).

As shown in Figure 1-1, the jig consists of a ring with four lugs for clamping and is frequently called a *ring jig*. It is attached to the cylinder head and held by U-bolt clamps. When used as a

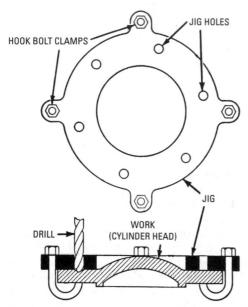

Figure 1-1   A plain ring-type clamp jig without bushings.

guide for the drill in the drilling operation, the jig makes certain that the holes are in the correct locations because the holes in the jig were located originally with precision. Therefore, laying out is not necessary.

A disadvantage of the simple clamp jig is that only holes of a single size can be drilled. Either *fixed* or *removable* bushings can be used to overcome this disadvantage. Fixed bushings are sometimes used because they are made of hardened steel, which reduces wear. Removable bushings are used when drills of different sizes are to be used, or when the drilled holes are to be finished by reaming or tapping.

A *bushed clamp jig* is illustrated in Figure 1-2. In drilling a hole for a stud, it is evident that the drill (tap drill) must be smaller in size than the diameter of the stud. Accordingly, two sizes of twist drills are required in drilling holes for studs. The smaller drill (or *tap drill*) and a drill slightly larger than the diameter of the stud are required for drilling the holes in the cylinder head. A bushing can be used to guide the tap drill.

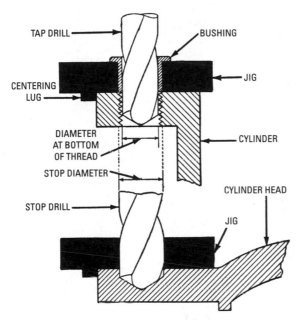

Figure 1-2 A clamp jig, with the tap drill guided by a bushing, designed for drilling holes in the cylinder (top); the operation for a hole for the cylinder head (bottom).

The jig is clamped to the work after it has been centered on the cylinder and head so that the axes of the holes register correctly. Various provisions (such as stops) are used to aid in centering the jig correctly. The jig shown in Figure 1-2 is constructed with four lugs as a part of the jig. As the jig is machined, the inner sides of the lugs are turned to a diameter that will permit the lugs to barely slip over the flange when the jig is applied to the work.

A *reversible clamp jig* is shown in Figure 1-3. The distinguishing feature of this type of jig is the method of centering the jig on the cylinder and head. The position of the jig for drilling the cylinder is shown at the top of Figure 1-3. An annular projection on the jig fits closely into the counterbore of the cylinder to locate the jig concentrically with the cylinder bore.

The jig is reversed for drilling the cylinder head. That is, the opposite side is placed so that the counterbore or circular recessed part of the jig fits over the annular projection of the cylinder head at the bottom of Figure 1-3.

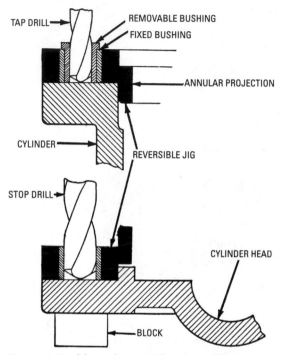

Figure 1-3 Note the use of a reversible clamp jig for the tap drill operation (top), and reversing the jig to drill the hole for the stud in the cylinder head (bottom).

This type of jig is often held in position by inserting an accurately fitted pin through the jig and into the first hole drilled. The pin prevents the jig from turning with respect to the cylinder as other holes are drilled.

A simple jig that has locating screws for positioning the work is shown in Figure 1-4. The locating screws are placed in such a way that the clamping points are opposite the bearing points on the work. Two setscrews are used on the long side of the work, but in this instance, because the work is relatively short and stiff, a single lug and setscrew (*B* in Figure 1-4) is sufficient.

This is frequently called a *plate jig* since it usually consists of only a plate that contains the drill bushings and a simple means of clamping the work in the jig, or the jig to the work. Where the jig is clamped to the work, it sometimes is called a *clamp-on jig*.

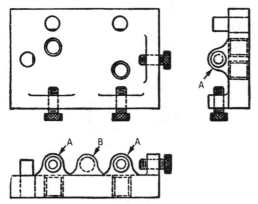

Figure 1-4   A simple jig that uses locating screws to position the work.

Diameter jigs provide a simple means of locating a drilled hole exactly on a diameter of a cylindrical or spherical piece (Figure 1-5).

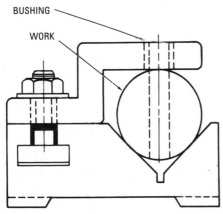

Figure 1-5   Diameter jig.

Another simple clamp jig is called a *channel jig* and derives its name from the cross-sectional shape of the main member, as shown in Figure 1-6. They can be used only with parts having fairly simple shapes.

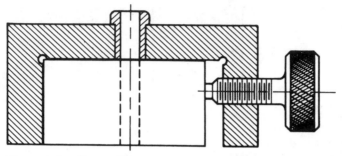

Figure 1-6 Channel jig.

Box Jig

Box jigs (sometimes called *closed jigs*) usually resemble a boxlike structure. They can be used where holes are to be drilled in the work at various angles. Figure 1-7 shows a design of box jig that is suitable for drilling the required holes in an engine link. The jig is built in the form of a partly open slot in which the link is moved up against a stop and then clamped with the clamp bolts A, B, and C.

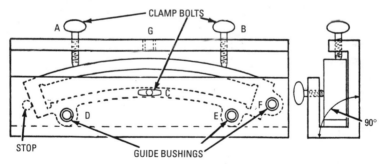

Figure 1-7 Using the box jig for drilling holes in an engine link.

The bushings *D* and *E* guide the drill for drilling the eccentric rod connections, and the bushing *F* guides the drill for the reach rod connections. The final hole, the hole for lubrication at the top of the link, is drilled by turning the jig 90°, placing the drill in the bushing *G*.

This type of jig is relatively expensive to make by machining, but the cost can be reduced by welding construction, using plate metal. In production work, the pieces can be set and released quickly.

A box jig with a hinged cover or leaf that may be opened to permit the work to be inserted and then closed to clamp the work into position is usually called a *leaf jig* (Figure 1-8). Drill bushings are usually located in the leaf. However, bushings may be located in other surfaces to permit the jig to be used for drilling holes on more than one side of the work. Such a jig, which requires turning to permit work on more than one side, is known as a *rollover jig*.

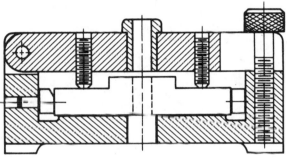

Figure 1-8 Leaf jig.

A box jig for angular drilling (Figure 1-9) is easily designed by providing the jig with legs of unequal length, thus tilting the jig to the desired angle. This type of jig is used where one or more holes are required to be drilled at an angle with the axis of the work.

As can be seen in Figure 1-9, the holes can be drilled in the work with the twist drill in a vertical position. Sometimes the jig is mounted on an angular stand rather than providing legs of unequal length for the jig. Figure 1-10 shows a box jig for drilling a hole in a ball.

In some instances, the work can be used as a jig (Figure 1-11). In the illustration, a bearing and cap are used to show how the work can be arranged and used as a jig. After the cap has been planed and fitted, the bolt holes in the cap are laid out and drilled. The cap is clamped in position, and the same twist drill used for the bolt holes is used to cut a conical spot in the base. This spotting operation provides a starting point for the smaller tap drill (*A* and *B* in Figure 1-11).

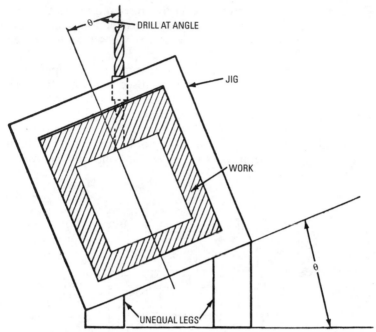

Figure 1-9 A box jig with legs of unequal length, used for drilling holes at an angle.

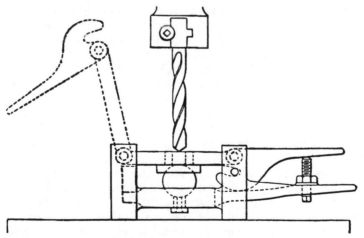

Figure 1-10 A box jig used for drilling a hole in a ball.

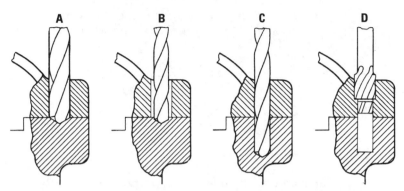

Figure 1-11 Using the work as a jig. In (A) the same drill used for the bolt holes is used to cut a conical spot in the base. This forms a starting point for the smaller tap drill, as shown in (B). In (C), the cap and bearing are clamped together and drilled by means of a tap drill, after which the tap drill is removed and a counterbore is used to enlarge the holes for the bolts, as shown in (D).

Also, both parts can be clamped together and drilled with a tap drill (C in Figure 1-11). Then, the tap drill can be removed and the holes for the bolts enlarged by means of a counterbore (D in Figure 1-11).

Following are some factors of prime importance to keep in mind with jigs:

- Proper clamping of the work
- Support of the work while machining
- Provision for chip clearance

When excessive pressure is used in clamping, some distortion can result. If the distortion is measurable, the result is inaccuracy in final dimensions. This is illustrated in an exaggerated way in Figure 1-12. The clamping forces should be applied in such a way that will not produce objectionable distortion.

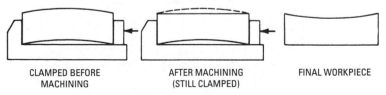

CLAMPED BEFORE MACHINING   AFTER MACHINING (STILL CLAMPED)   FINAL WORKPIECE

Figure 1-12 Effects of excessive pressure.

It is also important to design the clamping force in such a way that the work will remain in the desired position while machining, as shown in Figure 1-13.

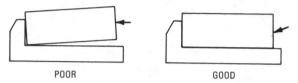

POOR   GOOD

Figure 1-13   Effects of clamping force.

Figure 1-14 shows the need for the jig to provide adequate support while the work is being machined. In the example shown in Figure 1-12, the cutting force should always act against a fixed portion and not against a movable section. Figure 1-13 illustrates the need to keep the points of clamping as nearly as possible in line with the cutting forces of the tool. This will reduce the tendency of these forces to pull the work from the clamping jaws. Support beneath the work is necessary to prevent the piece from distorting. Such distortion can result in inaccuracy and possibly a broken tool.

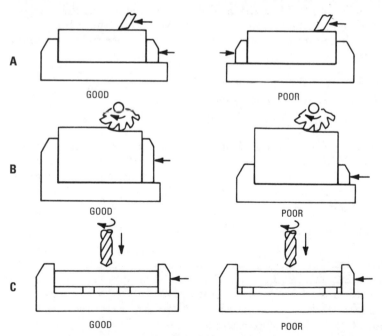

A

GOOD   POOR

B

GOOD   POOR

C

GOOD   POOR

Figure 1-14   Support for work during machining.

Adequate provision must be made for chip clearance, as illustrated in Figure 1-15. The first problem is to prevent the chips from becoming packed around the tool. This could result in overheating and possible tool breakage. If the clearance is not great enough, the chips cannot flow away. If there is too much clearance, the bushing will not guide the tool properly.

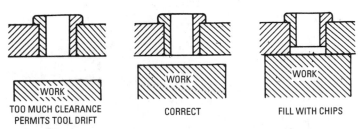

Figure 1-15   Provision for chip clearance.

The second factor in chip clearance is to prevent the chips from interfering with the proper seating of the work in the jig, as shown in Figure 1-16.

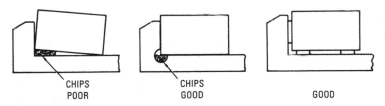

Figure 1-16   Provision for chip clearance.

Fixtures

As mentioned previously, a fixture is primarily a *holding* device. A fixture anchors the workpiece firmly in place for the machining operation, but it does not form a guide for the tool.

It is sometimes difficult to differentiate between a jig and a fixture, since their basic functions can overlap in the more complicated designs. The best means of differentiating between the two devices is to apply the basic definitions, as follows:

- The jig is a *guiding* device.
- The fixture is a *holding* device.

A typical example of a fixture is the device designed to hold two or more locomotive cylinders in position for planing (Figure 1-17). This fixture is used in planing the saddle surfaces. In the planing operation, two or more cylinders are placed in a single row, the fixture anchoring them firmly to the planer bed.

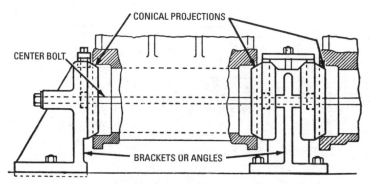

Figure 1-17   A fixture used to hold locomotive cylinders in position for planing the surfaces of the saddles.

The fixture consists of heavy brackets or angles, with conical projections that permit the bores of the cylinders to be aligned accurately with each other. The end brackets are made with a single conical flange; the intermediate brackets are made with double conical flanges. A bolt through the center of the flanges aligns the cylinder bores when it is tightened. The legs of the 90°-angle brackets at the ends are bolted firmly to the planer table. The intermediate brackets are also bolted to the planer table and aid in holding the assembly in firm alignment for the machining operation. The use of fixtures can result in a considerable saving in the time required to set the work, and they also ensure production of accurate work.

An *indexing fixture* can be used for machining operations that are to be performed in more than one plane (Figure 1-18). It facilitates location of the given angle with a degree of precision.

A disc in the indexing fixture is held in angular position by a pin that fits into a finished hole in the angle iron and into one of the holes in the disc. The disc is clamped against the knee by a screw and washer while the cut is being taken. Since the holes are properly spaced in the disc (index plate), the work attached to the disc can be rotated into any desired angular position. Radial drilling

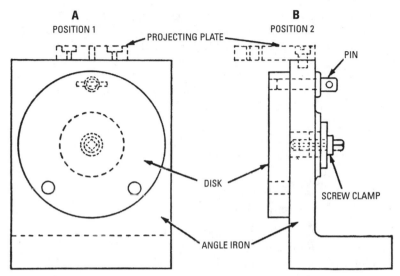

Figure 1-18   A simple type of indexing fixture that can be used to facilitate machining at accurately spaced angles.

operations can be performed when a projecting plate is provided with a jig hole.

The same general principles concerning clamping, support while machining, and chip clearance as covered in jigs apply as well to fixtures.

Summary

Jigs and fixtures are devices used to locate and hold the work that is to be machined. A jig is a guiding device, and a fixture is a holding device. A jig or fixture can be designed for a particular job. The form to be used depends on the shape and requirements of the workpiece that is to be machined.

There are generally two types of jigs used: the clamp jig and the box jig. Various names are applied to jigs (such as drilling, reaming, and tapping) according to the operation to be performed. Clamp jigs are sometimes called open jigs. Frequently, jigs are named for their shape, such as plate, ring, channel, and leaf.

A fixture anchors the workpiece firmly in place for the machining operation, but it does not form a guide for the tool. It is sometimes difficult to differentiate between a jig and a fixture, since their basic functions can overlap in the more complicated designs.

A plate jig consists of a plate, which contains the drill bushings, and a simple means of clamping the work in the jig, or the jig to the work. Where the jig is clamped to the work, it sometimes is called a clamp-on jig.

An indexing fixture can be used for machining operations that are to be performed in more than one plane. It facilitates location of the given angle with a degree of precision.

Review Questions

1. What are jigs and fixtures?
2. What does a jig do?
3. What is another name for a clamp jig?
4. What is the purpose of a fixture?
5. What is the disadvantage of a simple clamp jig?
6. What is another name for a box jig?
7. What can excessive jig pressure do?
8. What is an indexing fixture used for?
9. The fixture is primarily a _____ device.
10. The jig is primarily a _____ device.

Chapter 2

Helix and Spiral Calculations

In the past, machinists have tended to use the terms *helix* and *spiral* interchangeably. Generally, in machine shop usage, the terms should not be used interchangeably. These terms should be understood by machinists, and their misuse should be avoided.

For general machine shop usage, the terms can be defined as follows:

- A *helix* is a curve generated from a point that both rotates and advances axially on a cylindrical surface. The lead screw on a lathe is an example of a helix.
- A *spiral* is a curve generated from a point that has three distinctive motions: (a) rotation about the axis; (b) advancement parallel with the axis; and (c) an increasing or decreasing distance (radius) from the axis.

When a *cylindrical* workpiece is placed between centers on a milling machine and rotated by the index head as the table advances, a *helical groove* is milled by the cutter. When a *tapered* workpiece is placed between centers, tilted so that the top element is horizontal and then rotated by the dividing head as the table advances, a *spiral groove* is milled by the cutter. The basic difference between a helix and a spiral is illustrated in Figure 2-1.

Milling a Helix

Following are the essential requirements for milling a helix:

- The table should be set at the correct angle.
- The index head should be set to rotate the work in correct ratio to the table movement.
- The work should be fed toward the cutter by the table movement.

The *pitch*, or *lead*, of a helix is the distance that the table (carrying the workpiece) travels as the work is rotated by the index head through one complete revolution (Figure 2-2). The terms *lead* and *pitch* are identical in meaning. *Pitch* is probably a more proper term; however, lead is more commonly used in the machine shop.

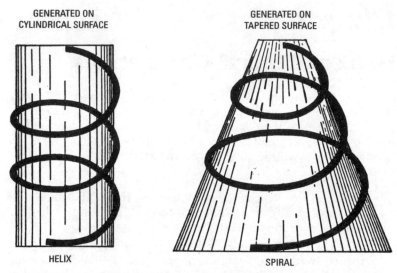

GENERATED ON
CYLINDRICAL SURFACE

GENERATED ON
TAPERED SURFACE

HELIX

SPIRAL

Figure 2-1   Basic difference between a helix (left) and a spiral (right).

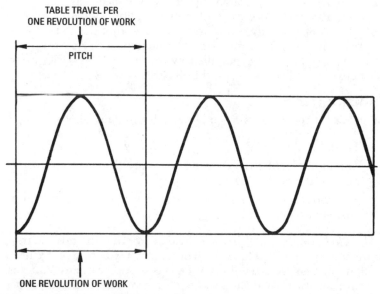

TABLE TRAVEL PER
ONE REVOLUTION OF WORK

PITCH

ONE REVOLUTION OF WORK

Figure 2-2   The pitch of a helix.

Angle of Table Swivel

This angle is the angle through which the table must be turned to cut a helix. The *table angle* is equal to the angle of the helix. Two methods can be used to determine the table angle for cutting a helix.

If a helix is laid out in a single plane, the hypotenuse of a right triangle represents the helix. The other two sides of the right triangle represent the circumference of the work and the pitch (Figure 2-3).

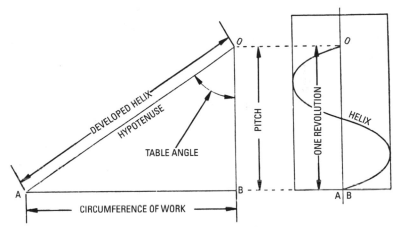

Figure 2-3   Development of a helix by laying out, to determine the table angle.

The angle *AOB* in Figure 2-3 (which is the angle of the helix) is called the *table angle* in the illustration because it is the angle through which the table must be turned to cut the helix correctly. If the triangle were cut out and wrapped around a cylindrical workpiece, the hypotenuse *OA*, which represents the developed helix, would coincide at all points with the helix.

The correct table position for cutting a helix is illustrated by angle *A* in Figure 2-4. Angle *A* is equal to angle *B*, which is called the *angle of the helix* and is formed by the intersection of the helix and a line parallel with the axis of the work. Angle *A* is equal to angle *B* because their corresponding sides are perpendicular. The helix angle depends on the pitch of the helix and the diameter of the work, and it varies inversely with the pitch for any given diameter.

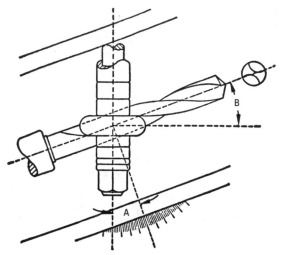

Figure 2-4   Correct position of the table for cutting a helix.

Turning the table to an angular position for cutting the helix prevents distortion of the shape of the cut and obtains clearance for the milling cutter. The pitch of the helix is not changed by turning the table to any angular position.

Trigonometry provides a more accurate method of determining the table angle. If the pitch and circumference of the work are given, the tangent of the table angle can be found. The pitch and circumference of the work are considered as the sides of a right triangle (Figure 2-5). After determining the value of the tangent, the angle can be obtained from a table of natural tangents.

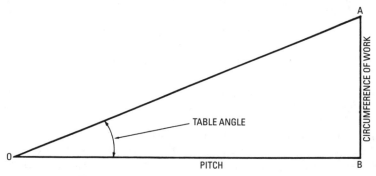

Figure 2-5   Using trigonometry to determine the table angle.

If, in the triangle AOB in Figure 2-5, we let the side AB equal the circumference of the work and let the side OB equal the pitch, then

$$\text{Tangent of the table angle} = \frac{AB \text{ (circumference)}}{OB \text{ (pitch)}}$$

For example, determine the table angle required to cut a helix that has a pitch of 16 inches and a diameter of 4.5 inches. The circumference of the work is equal to πD (3.141592654 × 4.5). Substituting in the formula,

$$\text{Tangent table angle} = \frac{3.141592654 \times 4.5 \text{ (circumference)}}{16 \text{ (pitch)}}$$

$$= 0.8835729338$$

Thus, the corresponding angle is approximately 41.46294384° (from the calculator's table of natural tangents).

If a protractor is used to measure the angle, it will be 41°27.8′ or 41°28′.

Lead of the Machine

To cut a helix or spiral, the table feed screw is connected to the spindle through a train of change gears. Therefore, for a given gear combination, the table advances a definite distance during each complete revolution of the spindle of the index head. If the change gears (which can be compared to those of the lathe) are all the same size, so that they do not change the velocity ratio between the table feed screw and the index head spindle, the table travel (in inches) per revolution of the index head spindle is the *lead of the machine*, which is identical to the *pitch of the machine*. The selection of the correct combination of change gears is important.

Thus, if the velocity ratio between the table feed screw and the index head spindle is unchanged, the cutter will mill a helix that has a pitch equal to the lead of the machine. If it is desirable to mill a helix that has a pitch different from the lead of the machine, change gears can be interposed to change the velocity ratio, so that a helix of the desired pitch can be produced.

If the lead, or pitch, of the table feed screw is ¼ inch (four threads per inch), the worm rotates 40 turns to one turn of the worm wheel, which is attached to the index head spindle. Thus, the change gears all have the same diameter, and the lead of the machine is the standard 10 inches (40 × ¼). This means that the table advances a distance of ¼ inch per revolution of the table feed screw. As the table feed screw makes 40 revolutions to one revolution of the index head spindle, the lead of the machine is 10 inches.

To mill a helix that has a pitch less than 10 inches, change gears that increase the speed of the worm shaft must be interposed; decrease the speed of the worm shaft for a pitch greater than 10 inches.

Change Gears

The corresponding velocity ratios must be calculated for the different pitches that can be used to mill the various kinds of helix. To meet these requirements, the change gears can be arranged as simple gearing and compound gearing.

For *simple gearing*, it is necessary only to select change gears that change the velocity ratio as follows:

$$\text{Velocity ratio} = \frac{\text{pitch of helix}}{\text{lead of machine}}$$

The change gears can be selected most conveniently by increasing both terms of the ratio to correspond with the number of teeth of the change gears available. When the change gears cannot be selected in this manner, compound gearing must be used.

Change-Gear Train

The change-speed gear train is composed of four gears as follows:

- The gear on the table feed screw shaft
- The first stud gear, so called because it is the first gear to be positioned
- The second stud gear
- The gear on the worm

The gear on the worm is somewhat a misnomer, as it is not actually the gear on the worm. It is the gear on a shaft that has a bevel gear on the opposite end of the shaft that meshes with another bevel gear of the same size on the worm shaft. For lack of a better name and because there is no change in the velocity ratio, the result is equivalent to its being placed directly on the worm shaft.

The four gears in the gear train are illustrated in Figure 2-6 and Figure 2-7. The gear on the table screw shaft and the first stud gear are the *driver* gears; the second stud gear and the gear on the worm are the *driven* gears.

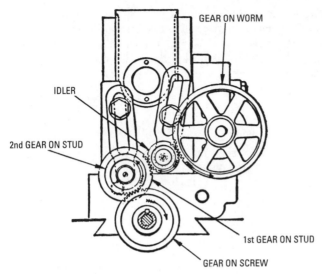

Figure 2-6 Diagram of change gearing, showing the use of the idler.

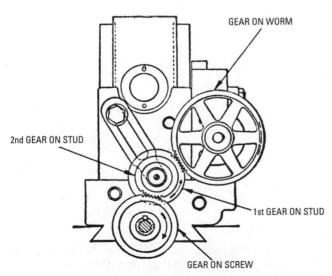

Figure 2-7 Change gearing that requires no idler.

Change-Gear Ratio

Different combinations of change gears can be used to determine the distance that the table moves during one revolution of the spindle. The pitch of the helix to be milled depends on the change-gear ratio. Expressed as a formula, the change-gear ratio can be determined as follows:

$$\text{change-gear ratio} = \frac{\text{pitch of helix}}{\text{lead of the machine}}$$

For example, if the lead of the machine is 10 inches, what change-gear ratio is required to cut a helix that has a pitch of 10.50 inches? Substituting in the formula,

$$\text{Change-gear ratio} = \frac{10.5}{10}$$

Multiply by 10 to get

$$\frac{105}{100}$$

Then, divide by 5, top and bottom, to get

$$\frac{21}{20}$$

or multiply this by 2 to produce

$$\frac{42}{40}$$

Therefore, a 20-tooth gear is placed on the table feed screw, and a 21-tooth gear is placed on the worm-shaft extension. If connected by idlers, the combination would provide the 10.5 pitch for the helix. However, there are no gears on the list of available gears that have 20 or 21 teeth. A 40-tooth gear is available in the set, but a 42-tooth gear is not available. Thus, calculations for equivalent gears must be made.

Change-Gear Calculations

Basically, these calculations are the same as for the change gears of an engine lathe. If change gears having the same diameter are used, a helix that has a pitch equal to the lead of the machine (standard lead is 10 inches) will be produced.

Equations illustrating the relationship of the different values are

$$\text{Change-gear ratio} = \frac{\text{pitch of helix}}{\text{lead of machine}}$$

$$= \frac{\text{pitch of helix}}{10 \text{ (standard lead)}}$$

and

$$\frac{\text{Driven gears}}{\text{Driver gears}} = \frac{\text{pitch of helix}}{10}$$

Since the product of each kind of gear determines the change-gear ratio,

$$\frac{\text{Product of driven gears}}{\text{Product of driver gears}} = \frac{\text{pitch of required helix}}{10}$$

The compound ratio of the driven gears can always be represented by a fraction. The numerator indicates the pitch to be cut, and the denominator indicates the lead of the machine. For example, if the required pitch is 20 inches and the lead of the machine is 10 inches (standard), the ratio is 20:10. Expressed in units, the ratio is the same as one-tenth of the required pitch to one. A convenient means of remembering the ratio is as follows: If the pitch is 40, the ratio of the gears is 4:1; if the pitch is 25, the ratio is 2.5:1; and so on.

As an example, determine the necessary gears to be used in milling a helix that requires a 12-inch pitch.

The compound ratio of the driven to the driver gears is as follows:

$$\frac{\text{Product of driven gears}}{\text{Product of driver gears}} = \frac{\text{pitch of required helix}}{10} = \frac{12}{10}$$

This fraction can be resolved into factors to represent the two kinds of change gears as follows:

$$\frac{12}{10} = \frac{(3 \times 4)}{(2 \times 5)}$$

Then, each term can be multiplied by a number common to both—24, in this instance—so that the numerator and denominator will correspond to the number of teeth of two change gears that are available with the machine. These multiplications do not affect the value of the fraction as shown here:

$$\frac{3 \times 24}{2 \times 24} = \frac{72}{48}$$

Likewise, the second pair of factors can be treated similarly:

$$\frac{4 \times 8}{5 \times 8} = \frac{32}{40}$$

Therefore, the driven gears (72 and 32) and the driver gears (48 and 40) are selected as follows:

$$\frac{12}{10} = \frac{\text{product of driven gears}}{\text{product of driver gears}} = \frac{72 \times 32}{48 \times 40}$$

As has been indicated, the first selected pair of gears (72 and 32) are the driven gears because the numerators of the fractions represent the driven gears (gear on worm and the second stud gear). Therefore, the 72-tooth gear is placed on the worm, and the 32-tooth gear is the second stud gear. The second pair of gears (48 and 40) are the driver gears because the denominators of the fractions represent the driver gears (gear on table feed screw and the first stud gear). Therefore, the 48-tooth gear is placed on the table feed shaft, and the 40-tooth gear is the first stud gear.

The steps for determining the change gears required for cutting a helix having a given pitch can be summarized as follows:

1. Determine the ratio between the required pitch of the helix and the lead of the machine (10 is a standard lead).
2. Express the ratio in the form of a fraction.
3. Resolve the fraction into two factors.
4. Raise the factors to higher terms, so that they correspond to the number of teeth in gears that are available with the machine.
5. The numerators represent the driven gears (gear on worm and second stud gear).

6. The denominators represent the driver gears (gear on feed screw and first stud gear).

7. Add an idler gear to cut a left-hand helix (on most machines).

As an example, select the gears for cutting a helix with a pitch of 27 inches.

$$\frac{27}{10} = \frac{3}{2} \times \frac{9}{5} = \frac{3}{2} \times \frac{16}{16} \times \frac{9}{5} \times \frac{8}{8} = \frac{48 \times 72}{32 \times 40}$$

The gear on the worm and the second stud gear are the 48-tooth gear and the 72-tooth gear, respectively. The gear on the table feed screw and the first stud gear are the 32-tooth gear and the 40-tooth gear, respectively.

Change-gear calculations can be checked by multiplying the product of the driven gears (48 × 72) by 10 and dividing by the product of the driver gears (32 × 40). The quotient is equal to the pitch of the resulting helix:

$$\frac{48 \times 72}{32 \times 40} \times 10 = 27 \text{ inches pitch}$$

This check is derived from the fact that the quotient of the product of the driven gears divided by the product of the driver gears is equal to the pitch of the helix divided by 10 (standard lead of the machine), or one-tenth of the pitch. Thus, ten times the product of the driven gears divided by the product of the driver gears is equal to the pitch of the helix.

Milling a Spiral

When tapered reamers, bevel gears, and so on, are to be held between centers and milled, that is, the cuts are to be taken at an angle to the axis of the work, the axis of the index head and the tailstock center should coincide with the axis of the work. If they do not coincide, errors in indexing and problems in machining are introduced. A typical setup for milling tapered work is shown in Figure 2-8.

A tilting table, an adjustable tailstock, or a taper attachment can be used to mill a piece of work that is tapered. These devices aid in mounting the work correctly (Figure 2-9).

The taper attachment (Figure 2-9) has one end attached to the spindle. The opposite end is bolted to a slotted bracket that is mounted on the table as shown in the diagram. If neither the tilting table nor the taper attachment is available, several objectionable

Figure 2-8 Setup for milling tapered work on the milling machine.
(Courtesy Cincinnati Milacron Co.)

AXIS OF CENTER AND
AXIS OF WORK COINCIDE

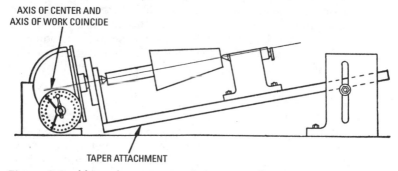

TAPER ATTACHMENT

Figure 2-9 Using the taper attachment to mill tapered work.

methods of mounting the tapered workpiece are often employed. Sometimes the tailstock is blocked up to the required height—with the index head having no angular adjustment (Figure 2-10). In this arrangement the work does not bear properly on the centers, and errors are introduced because of the angularity of the dog and the reciprocating motion of the tail of the dog in the slot of the driver. Misalignment of centers results in an uneven and wobbly bearing.

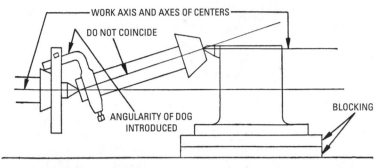

Figure 2-10   Using blocks to raise the tailstock for milling tapered work. This practice is objectionable but is used in the absence of a tilting table or taper attachment.

In milling machine work, there should be no lost motion between the tail of the dog and the driver plate. However, as shown in Figure 2-10, it is necessary to clamp the tailstock loosely to allow for the reciprocating motion of the tail of the dog, which is caused by the angularity of the dog. The angularity of the dog causes variation in the angular motion of the spindle and the work. Thus, indexing errors are introduced.

To index the work at an angle of 180° (*A* and *B* in Figure 2-11), the spindle would have to be indexed either more or less than 180°

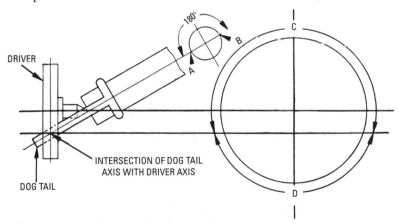

Figure 2-11   Note the variation of the rotation of the work. This is because of the angularity of the work and is caused by the angularity of the tail of the dog. If the head is indexed at 180°, rotation of the work is either more or less than 180°, depending on the direction of rotation.

(*C* and *D* in Figure 2-11), depending on the direction of rotation. This is because of the angularity of the dog.

Summary

A spiral is a curve generated from a point that has three distinctive motions. These three distinctive motions are rotation about the axis, advancement parallel with the axis, and increasing or decreasing distance from the axis or radius. A helix is a curve generated from a point that both rotates and advances axially on a cylindrical surface. The lead screw on a lathe is an example of a helix.

The pitch, or lead, of a helix is the distance the table travels as the work is rotated by the index head through a complete revolution. The two terms are identical in meaning. *Pitch* is probably a more proper term, but *lead* is more commonly used in the machine shop.

To cut a helix or spiral, the table feed screw is connected to the spindle through a train of change gears. When milling a helix, the table angle is equal to the angle of the helix. The table angle is the angle that the table must be turned to cut a helix. Two methods can be used to determine the table angle for cutting a helix.

A tilting table, an adjustable tailstock, or a taper attachment can be used to mill a piece of work that is tapered.

Review Questions

1. What is a helix?
2. What is a spiral?
3. What are the three distinct motions of a spiral?
4. How is a helical groove milled?
5. What are the three essential requirements for milling a helix?
6. The _____ of a helix is the distance that the table travels as the work is rotated by the index head through one complete revolution.
7. The _____ angle is equal to the angle of the helix.
8. Describe the pitch of a helix.
9. What is a hypotenuse?
10. The lead of the machine is identical to the _____ of the machine.
11. What is the change-gear ratio?

12. Of what is the change-gear train composed?

13. A tilting table, an adjustable tailstock, or a taper _____ can be used to mill a piece of work that is tapered.

14. What is used to raise the tailstock for milling tapered work?

15. If the head is indexed at 180°, rotation of the work is either more or less than _____ degrees.

Chapter 3

Spur Gear Computations

A *gear* is a form of disc, or wheel, having teeth around its periphery for providing a positive drive by meshing these teeth with similar teeth on another gear or rack. The slipping action of a belt (or other drive depending on friction) is eliminated with such a gear arrangement. A small amount of lost motion or *backlash* occurs between any two connected gears, but this is taken up by movement of the driving gear to give positive drive. In any two connected gears, the gear that receives the power is the *driver gear*, and the gear to which the power is delivered is the *driven gear*.

Evolution of Gears

A smooth cylinder mounted on a shaft may be considered to be a gear having an infinite number of teeth. The fundamental principle of toothed gearing is illustrated by a pair of cylinders mounted on parallel shafts with their surfaces in contact and rolling together in opposite directions (Figure 3-1).

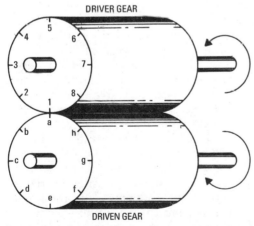

Figure 3-1   The evolution of gears. Two cylinders in rolling contact.

Because the teeth on a smooth cylinder are infinitely small, they do not project above the cylindrical surface. If power is applied to the driver cylinder in the direction of the arrow, the driven cylinder will turn by friction. This friction is equivalent to the meshing of infinitely small teeth. If the cylinders are of equal size, they will turn at the same speed—that is, the equally spaced divisions (1, 2, 3, and so on) will coincide as the two cylinders rotate. This is true only as long as the load on the driven cylinder is not large enough to cause slippage. In this particular instance, very little friction is present to prevent slippage of the cylinders (line *EL* in Figure 3-2).

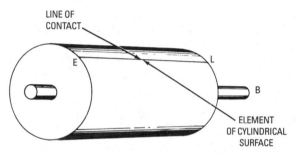

LINE OF
CONTACT

E

L

B

ELEMENT
OF CYLINDRICAL
SURFACE

Figure 3-2   The driven gear, showing the line of contact EL.

If the cylindrical surfaces were perfectly smooth, frictional contact would not be present to turn the cylinders. Although metal surfaces may appear smooth to the eye and to touch, minute irregularities are present even though they are invisible without magnification. Thus, the line of contact (line *EL* in Figure 3-3) has the appearance of a strip of coarse emery paper. When the two cylinders are in contact, the interlocking minute irregularities produce the frictional contact. When pressure is applied to force a firm contact, friction is increased by the flexing or flattening of the metal along the line of contact, thus increasing the contact area (Figure 3-4). This exaggerated illustration of flexing of surfaces may be compared to the action of a clothes wringer with its two rubber rolls under considerable pressure.

In the perfect surface, there are no minute irregularities or flexing. Therefore, there is no frictional contact, and the line of contact (line *EL* in Figure 3-3) is a part of the surface—that is, it has only one dimension (length), with no contact area. If machined surfaces were perfect surfaces, frictional contact would be impossible, and power could not be transmitted by cylinders. Hence, the necessity for toothed gears to obtain a positive drive is evident.

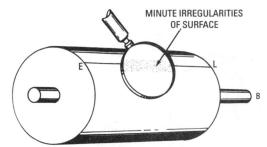

Figure 3-3 Note the minute irregularities of the surface in line of contact EL.

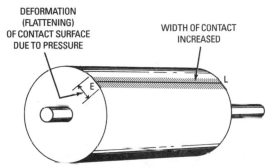

Figure 3-4 The width of the contact line EL is increased due to the flattening of the surface under pressure.

Gear Teeth

The position of the teeth with respect to the periphery of the cylinders should be understood before considering the various shapes of gear teeth and the method of generating these shapes. It should be noted that each tooth projects both above and below the periphery of the cylinder (Figure 3-5). If the surfaces of two cylinders are to remain in contact, teeth could not be formed in their surfaces by cutting grooves in them because it would be necessary to move their axes closer together for the teeth to interlock. Thus, the original surfaces would overlap. Teeth could not be added to the cylinders because the axes would have to be moved farther apart, thus separating the contact surfaces.

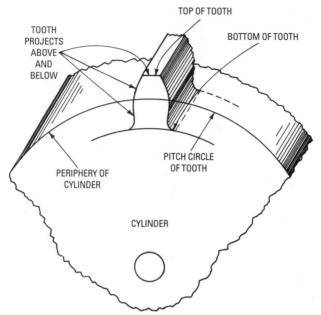

Figure 3-5 Position of gear tooth with respect to the periphery of the cylinder.

Therefore, a combination of the two methods must be used: cutting grooves equal to one-half the proposed depth plus clearance, and adding an equal amount between the spaces formed to complete the partly formed teeth. Then, the teeth will fall into the spaces and interlock properly, the original surfaces of the cylinders will remain on the contact line, and the diameters of the cylinders will provide the main circles for all calculations for speed, numbers, teeth dimensions, and so on.

Gear Tooth Terms

Of course, if the cylinders were gear blanks on which gear teeth were to be cut, the teeth could not project above the cylindrical surfaces of the blanks, but this is done in Figure 3-5 for purposes of illustration.

Pitch Circle

As mentioned, the original surfaces of the cylinders remain on the contact line, and the diameters of these cylinders (or circles) provide the basis for the various gear tooth computations. The *pitch circle* is the line of contact of the two cylinders. The pitch circle is the reference circle of measurement and is located one-half the distance between the top and bottom of the theoretical tooth (Figure 3-6).

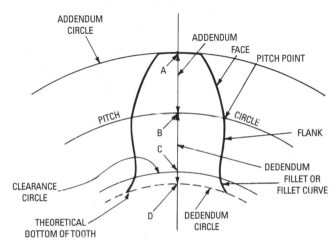

Figure 3-6   Diagram of a theoretical gear tooth.

The *pitch point* is the point of tangency of two pitch circles—or of a pitch circle and a pitch line—and is located on the line of centers. The point of intersection of a tooth profile with the pitch circle is its *pitch point.*

The *face* of a gear tooth is the surface of the tooth between the pitch circle and the top of the tooth. The *flank* of the tooth is the surface of the tooth between the pitch circle and the bottom of the groove, including the *fillet.* The fillet is a small arc (or *fillet curve*) that joins the tooth profile to the bottom of the tooth space, thus avoiding sharp corners at the root of the tooth.

Addendum Circle and Dedendum Circle
The *addendum circle* is the circle that passes through the top of the gear teeth. The diameter of the addendum circle is the same as the *outside diameter* of the gear. The *addendum* is the height of the tooth above the pitch circle, or the radial distance between the pitch circle and the top of the tooth. The circle that passes through the bottom of the tooth space is called the *dedendum circle.* The *dedendum* is the depth of the tooth space below the pitch circle, or the radial dimension between the pitch circle and the bottom of the tooth space (see Figure 3-6).

Spur Gear Computations

The spur gear is the simplest gear and the one in most common use. A spur gear has straight teeth cut parallel with the axis of rotation of the gear body. All the other gear forms—bevel gears, helical gears, worm gears, and worm wheels (Figure 3-7)—are modifications of the spur gear. The general principle, or the principle on which gear teeth are formed, is practically the same in all the forms of gears in use.

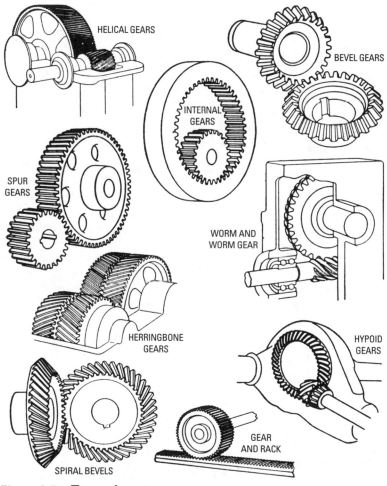

Figure 3-7 Types of gears.

Circular Pitch

The circular pitch is the distance from the center of one tooth to the center of the adjacent tooth as measured on the pitch circle. It may be considered as the width of one tooth plus the width of one space as measured on the pitch circle. Thus, the circular pitch is an arc whose length depends on the number of teeth in the gear and on the diameter of the pitch circle. In Figure 3-8 it may be noted that the circular pitch is equal to the length of the arc *ABC*. This arc is equal in length to the length of the arc *EF*, also on the pitch circle. Circular pitch may be determined by dividing the circumference of the pitch circle (πD) by the number of teeth as follows:

$$\text{Circular pitch} = \frac{\text{circumference of pitch circle}}{\text{number of teeth}}$$

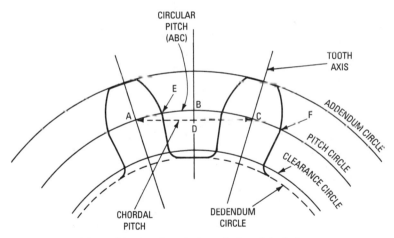

Figure 3-8 Note the circular pitch and chordal pitch.

If the circumference of the pitch circle and the circular pitch are known, the number of teeth in the gear may be calculated as follows:

$$\text{Number of teeth} = \frac{\text{circumference of pitch circle}}{\text{circular pitch}}$$

Likewise, if the circular pitch and the number of teeth in the gear are known, the diameter of the pitch circle may be calculated as follows:

$$\text{Diameter of pitch circle} = \frac{\text{circular pitch} \times \text{number of teeth}}{3.1416}$$

Chordal pitch is the distance from the center of one tooth to the center of another tooth when measured on the chord of an arc of the pitch circle (Figure 3-8).

The formulas for spur gears have been assembled together in Table 3-1 for convenience.

Table 3-1 Formulas for Spur Gear Calculations

No. of Teeth	Chordal Thickness	Chordal Addend.	No. of Teeth	Chordal Thickness	Chordal Addend.	No. of Teeth	Chordal Thickness	Chordal Addend.
10	1.56435	1.06156	59	1.57061	1.01046	108	1.57074	1.00570
11	1.56546	1.05598	60	1.57062	1.01029	109	1.57075	1.00565
12	1.56631	1.05133	61	1.57062	1.01011	110	1.57075	1.00560
13	1.56698	1.04739	62	1.57063	1.00994	111	1.57075	1.00556
14	1.56752	1.04401	63	1.57063	1.00978	112	1.57075	1.00551
15	1.56794	1.04109	64	1.57064	1.00963	113	1.57075	1.00546
16	1.56827	1.03852	65	1.57064	1.00947	114	1.57075	1.00541
17	1.56856	1.03625	66	1.57065	1.00933	115	1.57075	1.00537
18	1.56880	1.03425	67	1.57065	1.00920	116	1.57075	1.00533
19	1.56899	1.03244	68	1.57066	1.00907	117	1.57075	1.00529
20	1.56918	1.03083	69	1.57066	1.00893	118	1.57075	1.00524
21	1.56933	1.02936	70	1.57067	1.00880	119	1.57075	1.00519
22	1.56948	1.02803	71	1.57067	1.00867	120	1.57075	1.00515
23	1.56956	1.02681	72	1.57067	1.00855	121	1.57075	1.00511
24	1.56967	1.02569	73	1.57068	1.00843	122	1.57075	1.00507
25	1.56977	1.02466	74	1.57068	1.00832	123	1.57076	1.00503
26	1.56986	1.02371	75	1.57068	1.00821	124	1.57076	1.00499
27	1.56991	1.02284	76	1.57069	1.00810	125	1.57076	1.00495
28	1.56998	1.02202	77	1.57069	1.00799	126	1.57076	1.00491
29	1.57003	1.02127	78	1.57069	1.00789	127	1.57076	1.00487
30	1.57008	1.02055	79	1.57069	1.00780	128	1.57076	1.00483
31	1.57012	1.01990	80	1.57070	1.00772	129	1.57076	1.00479
32	1.57016	1.01926	81	1.57070	1.00762	130	1.57076	1.00475

(continued)

Table 3-1 (continued)

No. of Teeth	Chordal Thickness	Chordal Addend.	No. of Teeth	Chordal Thickness	Chordal Addend.	No. of Teeth	Chordal Thickness	Chordal Addend.
33	1.57019	1.01869	82	1.57070	1.00752	131	1.57076	1.00472
34	1.57021	1.01813	83	1.57070	1.00743	132	1.57076	1.00469
35	1.57025	1.01762	84	1.57071	1.00734	133	1.57076	1.00466
36	1.57028	1.01714	85	1.57071	1.00725	134	1.57076	1.00462
37	1.57032	1.01667	86	1.57071	1.00716	135	1.57076	1.00457
38	1.57035	1.01623	87	1.57071	1.00708	136	1.57076	1.00454
39	1.57037	1.01582	88	1.57071	1.00700	137	1.57076	1.00451
40	1.57039	1.01542	89	1.57072	1.00693	138	1.57076	1.00447
41	1.57041	1.01504	90	1.57072	1.00686	139	1.57076	1.00444
42	1.57043	1.01471	91	1.57072	1.00679	140	1.57076	1.00441
43	1.57045	1.01434	92	1.57072	1.00672	141	1.57076	1.00439
44	1.57047	1.01404	93	1.57072	1.00665	142	1.57076	1.00435
45	1.57048	1.01370	94	1.57072	1.00658	143	1.57076	1.00432
46	1.57050	1.01341	95	1.57073	1.00651	144	1.57076	1.00429
47	1.57051	1.01311	96	1.57073	1.00644	145	1.57077	1.00425
48	1.57052	1.01285	97	1.57073	1.00637	146	1.57077	1.00422
49	1.57053	1.01258	98	1.57073	1.00630	147	1.57077	1.00419
50	1.57054	1.01233	99	1.57073	1.00623	148	1.57077	1.00416
51	1.57055	1.01209	100	1.57073	1.00617	149	1.57077	1.00413
52	1.57056	1.01187	101	1.57074	1.00611	150	1.57077	1.00411
53	1.57057	1.01165	102	1.57074	1.00605	151	1.57077	1.00409
54	1.57058	1.01143	103	1.57074	1.00599	152	1.57077	1.00407
55	1.57058	1.01121	104	1.57074	1.00593	153	1.57077	1.00405
56	1.57059	1.01102	105	1.57074	1.00587	154	1.57077	1.00402
57	1.57060	1.01083	106	1.57074	1.00581	155	1.57077	1.00400
58	1.57061	1.01064	107	1.57074	1.00575	156	1.57077	1.00397

Diametral Pitch

The number of teeth in a gear per inch of pitch circle diameter is called the *diametral pitch*. The diametral pitch is a ratio of the number of teeth in a gear to the number of inches in the diameter of the pitch circle. Since it is a ratio between two quantities, diametral pitch cannot be shown on a blueprint as a dimension. The diametral pitch system is designed to designate a series of gear tooth sizes (whole numbers), just as screw-thread pitches are standardized to designate the size of the thread. So, it should be remembered that there must be a whole number of teeth in a gear. Diametral pitch is usually referred to as *pitch*. For example, a 10-pitch gear indicates that a gear has 10 teeth per inch of diameter of its pitch circle.

The diametral pitch of a gear may be determined by dividing the number of teeth in the gear by its pitch circle diameter:

$$\text{Diametral pitch} = \frac{\text{number of teeth}}{\text{pitch circle diameter}}$$

As an example, if a gear has 22 teeth, and the diameter of the pitch circle is 2 inches, calculate the diametral pitch as follows:

$$\text{Diametral pitch} = \frac{22}{2} = 11$$

The following relationships may be obtained from the equation:

1. The number of teeth in the gear may be obtained by multiplying the diametral pitch by the diameter of the pitch circle:

   Number of teeth = diametral pitch × diameter of pitch circle

   For example, given a diametral pitch of 11 and a pitch circle diameter of 2, you could find the number of teeth by using the following equation:

   Number of teeth = 11 × 2 = 22 teeth

2. The diameter of the pitch circle may be obtained by dividing the number of teeth by the diametral pitch as follows:

   $$\text{Diameter of pitch circle} = \frac{\text{number of teeth}}{\text{Diametral pitch}}$$

   For example, given that a gear has 22 teeth and a diametral pitch of 11, you could find the diameter of the pitch circle by using the following equation:

   $$\text{Diameter of pitch circle} = \frac{22}{11} = 2 \text{ inches}$$

Nearly all gear calculations are made in terms of diametral pitch rather than circular pitch because diametral pitch may usually be expressed as a whole number, which is convenient for expressing the proportions of teeth. A series of symbols and formulas may be used to determine the proportions of gear teeth (Figure 3-9).

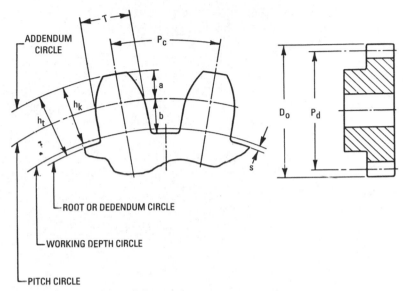

Figure 3-9 Diagram showing the proportions of spur gear teeth. These symbols are used in formulas for determining the proportion of spur gear teeth. *(Courtesy Cincinnati Milacron Co.)*

Frequently, it is convenient to change diametral pitch (P) to circular pitch (Pc), or vice versa. If either of the two terms is known, the other may be determined, as shown by the following rules and formulas:

1. To calculate diametral pitch (P) when circular pitch is known, divide 3.1416 by the circular pitch (Pc):

$$P = \frac{3.1416}{Pc}$$

As an example, if the circular pitch (Pc) of a gear is ⅝ (0.625) inch, what is the diametral pitch (P) to the nearest whole number?

$$P = \frac{3.1416}{0.625}$$

$$= 3.1416 \div 0.625$$

$$= 5.026 \text{ or } 5$$

2. To calculate circular pitch (*Pc*) when diametral pitch is known, divide 3.1416 by the diametral pitch (*P*):

$$Pc = \frac{3.1416}{P}$$

As an example, if the diametral pitch (*P*) of a gear is 5, what is the corresponding circular pitch (*Pc*)?

$$Pc = \frac{3.1416}{5} = 0.6283 \text{ or } \tfrac{5}{8} \text{ inch}$$

Pitch Diameter

The diameter of the pitch circle is called pitch diameter (*Pd*). The operating pitch diameter is the pitch diameter at which the gears operate. Pitch diameter (*Pd*) can be determined by dividing the number of teeth in the gear (*N*) by the diametral pitch (*P*) as follows:

$$Pd = \frac{N}{P}$$

As an example, if the gear has 22 teeth (*N*) and the diametral pitch is 11, what is the pitch diameter (*Pd*)?

$$Pd = \frac{22}{11} = 2 \text{ inches}$$

Addendum

The height by which a tooth projects above the pitch circle is called the addendum. It is also the radial distance between the pitch circle and the addendum circle (see Figure 3-9). In spur gear calculations, the addendum (*a*) is equal to 1.0 divided by the diametral pitch (*P*):

$$a = \frac{1.0}{P}$$

As an example, if the diametral pitch (*P*) of a gear is 11, what is the addendum (*a*), or height above the pitch circle?

$$a = \frac{1.0}{11} = 0.0909 \text{ inch}$$

Dedendum

The depth of a tooth space below the pitch circle is called the dedendum. It is the radial distance between the pitch circle and the root or dedendum circle (Figure 3-8). The dedendum (*b*) is calculated by dividing 1.157 by the diametral pitch (*P*):

$$b = \frac{1.157}{P}$$

As an example, calculate the dedendum (*b*) for a gear having a diametral pitch (*P*) of 11.

$$b = \frac{1.157}{11} = 0.1052 \text{ inch}$$

Number of Teeth in a Gear

Since the diametral pitch expresses the number of teeth per inch of pitch circle diameter, it follows that the number of teeth in a gear (*N*) can be found by multiplying the diameter of the pitch circle (*Pd*) by the diametral pitch (*P*) as follows:

$$N = Pd \times P$$

As an example, if the diametral pitch (*P*) of a gear is 10 and the pitch circle diameter (*Pd*) is 2 inches, how many teeth (*N*) are in the gear?

$$N = 2 \times 10 = 20 \text{ teeth}$$

Likewise, the diameter of the pitch circle (*Pd*) may be found by dividing the number of teeth (*N*) in the gear by the diametral pitch (*P*):

$$Pd = \frac{N}{P}$$

As an example, if there are 20 teeth in a gear (*N*) and the diametral pitch (*P*) is 10, what is the pitch diameter (*Pd*)?

$$Pd = \frac{20}{10} = 2 \text{ inches}$$

Clearance

As mentioned previously, gears must be proportioned in such a way that when a tooth is in mesh with the mating gear, the top of the tooth should not touch the bottom of the groove or space between the tooth and the adjacent tooth. The *clearance*, or radial distance from the dedendum circle to the clearance circle (Figure 3-9), provides a margin of space to allow for any slight errors in machining. The clearance is the amount by which the dedendum exceeds the addendum of a given gear. In actual practice, the clearance is equal to one-twentieth ($\frac{1}{20}$) of the circular pitch. In gear calculations, clearance (S) may be determined by dividing 0.157 (obtained by taking $\frac{1}{20}$ of 3.1416) by the diametral pitch (P), as shown by the following formula:

$$S = \frac{0.157}{P}$$

For example, calculate the clearance (S) for a gear having a diametral pitch (P) of 11.

$$S = \frac{0.157}{11} = 0.0143 \text{ inch clearance}$$

Working Depth
The theoretical length of the tooth is the *working depth*. The working depth of a tooth is equal to twice the addendum (see Figure 3-9). For two mating gears, the working depth is the sum of their addenda or the depth of engagement of the two gears. The working depth (h_k) of a gear may be calculated by dividing 2.0 by the diametral pitch (P), as shown by the following formula:

$$h_k = \frac{2.0}{P}$$

For example, if the diametral pitch (P) of a gear is 11, determine the working depth (h_k).

$$h_k = \frac{2.0}{11} = 0.1818 \text{ inch}$$

Whole Depth
This is the total depth of a tooth space and is equal to the addendum plus the dedendum. The *whole depth* is also equal to the working depth plus clearance (Figure 3-9). The whole depth (h_t) can be determined by dividing 2.157 by the diametral pitch (P), as shown in the following formula:

$$h_t = \frac{2.157}{P}$$

For example, if the diametral pitch (P) of a gear is 11, find the whole depth (h_t).

$$h_t = \frac{2.157}{11} = 0.196 \text{ inch}$$

Outside Diameter

The diameter of an addendum circle (Figure 3-9) is the *outside diameter* (D_o) of a gear. As the height of the teeth above the pitch circle is equal to 1 divided by the diametral pitch, the formula for outside diameter can be expressed as follows:

$$Do = Pd + \frac{1}{P} + \frac{1}{P}$$

A 20-tooth, 10-pitch gear would have a height above the pitch circle equal to $\frac{1}{10}$, or 0.1 inch (Figure 3-10). The diameter of the pitch circle is equal to $\frac{20}{10}$, or 2 inches. Thus, the outside diameter of the gear is equal to $0.1 + 0.1 + 2.0 = 2.2$ inches.

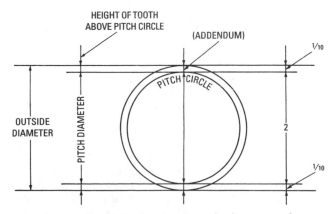

Figure 3-10   Diagram showing outside diameter of gear teeth.

The outside diameter of the gear (2.2 inches) is equal to the pitch diameter of a gear with two additional teeth (22 divided by 10 = $\frac{22}{10}$ = 2.2 inches). Therefore, the outside diameter (D_o) of a gear may be calculated by adding 2 to the number of teeth (N) and dividing the sum by the diametral pitch (P):

$$Do = \frac{N}{P} + \frac{1}{P} + \frac{1}{P}$$

$$Do = \frac{N + 2}{P}$$

For example, if a gear has 20 teeth (N) and the diametral pitch (P) is 10, find the outside diameter (D_o) of the tooth.

$$Do = \frac{20+2}{10} = \frac{22}{10} = 2.2 \text{ inches}$$

Tooth Thickness
This is the length of an arc of the pitch circle between the two sides of a gear tooth. *Tooth thickness* (T) may be calculated by dividing 1.5708 by the diametral pitch (P):

$$T = \frac{1.5708}{P}$$

For example, find the tooth thickness (T) if the diametral pitch (P) is 11.

$$T = \frac{1.5708}{11} = 0.1428 \text{ inch}$$

Side Clearance (Backlash)
Theoretically, the thickness of a tooth on the pitch circle is the same as the width of the groove. In practice, it is necessary to make the width of the space slightly larger than the thickness of the tooth to allow for inaccuracies of operation and workmanship. The difference in the thickness of the tooth and the width of the space is called *backlash*.

Center Distance
Since the diametral pitch designates the number of teeth in a gear per inch of diameter, the center distance (C) between two meshing gears can be determined by adding the number of teeth in both

gears ($N_1 + N_2$) and dividing the sum by two times the diametral pitch (P), as shown here:

$$C = \frac{(N_1 + N_2)}{2P}$$

As an example, if the diametral pitch (P) is 4, one gear has 60 teeth (N_1), and the other gear has 20 teeth (N_2), what is the distance between centers (C)?

$$C = \frac{60 + 20}{2(4)} = \frac{80}{8} = 10 \text{ inches}$$

Length of Rack
A rack is a gear with its teeth spaced along a straight line, making it suitable for straight-line motion. The length of a rack (L) may be determined by multiplying the number of teeth in the rack (N) by the circular pitch (Pc): $L = NPc$.

Pressure Angle
This is the angle between the tooth profile and a radial line at its pitch point. In involute teeth, the pressure angle is described as the angle between the line of action and the line tangent to the pitch circle (Figure 3-11).

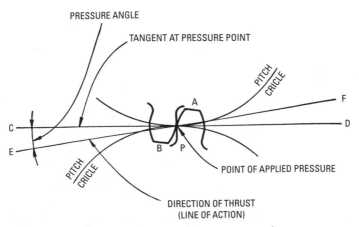

Figure 3-11   Diagram showing the pressure angle.

In Figure 3-11, the pitch circles of the two gears make contact at the pitch point P. Two mating teeth are shown in contact at that point on the pitch circles. The line *CD* is tangent to the pitch circles through pitch point P, and the line *EF* represents the direction of the applied pressure of the driving tooth (line of action). Angle *EPC*, which is between the line of action *EF* and the tangent line *CD*, is called the *pressure angle*.

In gear design, a pressure angle of 14.5° is most commonly used for involute teeth. It has been found by experience that maximum efficiency can be obtained through a pressure angle of 14.5°. The selection of the 14.5° pressure angle was originally influenced by the fact that the sine of 14.5° is approximately 0.25, a convenient proportion for the millwright to lay out. The involute stub tooth has a standard pressure angle of 20°. (Actually, the sine of 14.5° = 0.2503800041.)

In gear design, the pressure angles are either 14.5° or 20°. Formerly, it was believed that a pressure angle of 20° caused too much wear on the gear teeth and a rougher action between the gears. However, it has been found that the bearing pressure is not greater, wear on the teeth is no more, and the action is just as smooth for gears with a 20° pressure angle.

Involute Gears

The *involute curve* is used almost exclusively for gear tooth profiles. The form or shape of an involute curve is dependent on the diameter of the base circle from which it is derived.

An involute curve may be traced by a point on a taut string that is unwound from the circumference of a circle—the base circle of the involute. Figure 3-12 shows the generation of an involute curve.

To describe an involute of a given base circle, let *A, B, C*, and so on, be equal divisions on the circumference of the circle. At each of these points, draw tangents to the circle. On the tangent at *B*, make a distance *Bb* equal to the arc *AB*; on the tangent at *C*, make the distance *Cc* equal to two times arc *AB*, etc. Thus, the points *b, c, d*, and so on, are obtained. Through these points, describe the curve *A, b, c, d, e, f, g*, which is the involute of the base circle.

To illustrate the generation of the involute form of tooth (Figure 3-13), first describe the pitch circle, addendum circle, and dedendum circle with radii equal to the selected data for the tooth to be drawn. Select any point O as the pitch point on the

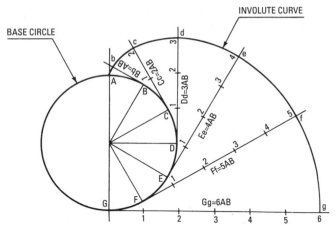

Figure 3-12 Diagram showing the method of describing an involute curve from a given base circle.

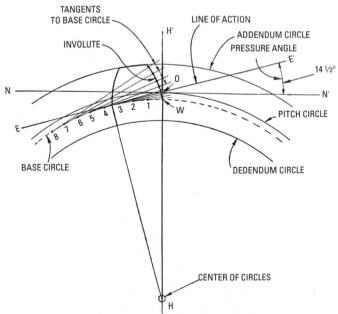

Figure 3-13 Diagram showing the method of generating an involute gear tooth.

pitch circle. Draw the line *HH'* from the center through the pitch point O, and also through point O draw line *NN'* at an angle of 90° to *HH'*. Line *NN'* is then tangent to the pitch circle at point O.

Through point O, draw the line of action *EE'*, at an angle of 14½° with line *NN'* (angle *E'ON'* = 14½°). This is the pressure angle generally used. With *H* as the center, describe the base circle with a radius that will make the base circle tangent to the line of action *EE'*. To obtain points on the involute curve, draw tangents to the base circle at the equally spaced points 1, 2, 3, 4, and so on, and locate the points on the involute, as shown in Figure 3-12.

It should be noted that the base circle is always smaller than the pitch circle. The involute curve forms the addendum of the tooth and extends to the base circle. The balance of the tooth (or flank) is drawn radially from the base circle, except for a small fillet at the bottom of the tooth.

Summary

A gear is a form of disc, or wheel, having teeth around its periphery for providing a positive drive. In any two connecting gears, the gear that receives the power is the driver gear, and the gear to which the power is delivered is the driven gear.

The positions of gears with respect to the periphery of the cylinders should be understood before considering the various shapes of gear teeth and the method of generating these shapes. Each tooth projects both above and below the periphery of the cylinder. If the surfaces of two cylinders are to remain in contact, teeth could not be formed in their surfaces by cutting grooves in them.

The spur gear is the simplest gear and the one most commonly used. The spur gear has straight teeth cut parallel with the axis of rotation of the gear body. All other gear forms, such as bevel, helical, and worm, are modifications of the spur gear.

The number of teeth in a gear per inch of pitch circle diameter is called diametral pitch. The diametral pitch is a ratio of the number of teeth in a gear to the number of inches in the diameter of the pitch circle. The diametral pitch of a gear may be determined by dividing the number of teeth in the gear by its pitch circle diameter. Nearly all gear calculations are made in terms of diametral pitch rather than circular pitch.

The height by which a tooth projects above the pitch circle is called the addendum. It is also the radial distance between the pitch circle and the addendum circle. In spur gear calculations, the addendum is equal to 1.0 divided by the diametral pitch. The depth of a tooth space below the pitch circle is called the dedendum. It is the radial distance between the pitch circle and the root or dedendum circle. The dedendum is calculated by dividing 1.157 by the diametral pitch.

Review Questions

1. What is a gear?
2. What is backlash?
3. What is the difference between the driver gear and the driven gear?
4. What is the line of contact?
5. What is meant by *deformation*?
6. What is pitch circle?
7. What is pitch point?
8. The _____ of a gear tooth is the surface of the tooth between the pitch circle and the top of the tooth.
9. What is meant by the *flank* of the tooth?
10. The diameter of the addendum circle is the same as the _____ diameter of the gear.
11. What is the difference between the addendum and the dedendum?
12. What is the circular pitch of a gear?
13. What is the difference between chordal pitch and circular pitch?.
14. What is a gear and rack used for?
15. What is chordal pitch?
16. Identify diametral pitch.
17. Diametral pitch is equal to the number of teeth divided by the _____ circle diameter.
18. Nearly all gear calculations are made in terms of diametral pitch rather than___ pitch.
19. What is the value of pi?

20. The theoretical length of the gear tooth is the working
_____.

21. What is whole depth?

22. What is meant by the term *side clearance?*

23. What is meant by the term *length of rack?*

24. What is meant by the term *pressure angle?*

25. What is an involute curve?

Chapter 4

Gears and Gear Cutting

The *spur gear* is the simplest gear and the one most commonly used. A spur gear has straight teeth that are cut parallel with the axis of rotation of the gear body. All the other gear forms (bevel gears, helical gears, worm gears, and worm wheels) are modifications of the spur gear. The general principle, or the principle on which gear teeth are formed, is practically the same in all the forms of gears in use.

Development of Gear Teeth

The curves for gear tooth profiles should be constructed in such a way that a uniform velocity ratio is attained when two wheels are geared together. The common normal at the point of contact should always pass through the pitch point, which divides the line of centers in the inverse ratio of the angular velocities.

The shapes of gear teeth differ with the system used in generating them. The two systems in general use are *involute* and *cycloidal*.

The involute curve is used almost exclusively for gear tooth profiles. The term *involute* refers to the shape of the curve. The form or shape of an involute curve is dependent on the diameter of the base circle from which it is derived. The curve can be traced by a point on a taut string that is unwound from a circumference of a circle—the *base circle of the involute*.

When two involute curves are brought into contact on the line of centers, a pitch point is established. The pitch point determines the diameter of the pitch circle. It is only in the involute form of tooth that the diameter of the pitch circle can be a flexible dimension, which permits involute gears to operate successfully at center distances that vary slightly.

Formerly, the cycloidal system was used almost exclusively. In the cycloidal system of generating gear teeth, the tooth profile is a double curve consisting of an *epicycloidal face* and a *hypocycloidal flank*. The double curve is generated by rolling circles that roll on the pitch circle. The epicycloidal portion of the curve is generated by an outer rolling circle, and the hypocycloidal portion is generated by an inner rolling circle. The diameters of the rolling circles should not be greater than the radius of the pitch circle to avoid generating a gear tooth that is weak at the root. Also, the diameter of the rolling circle should not be less than one-half the radius of the pitch circle (Figure 4-1).

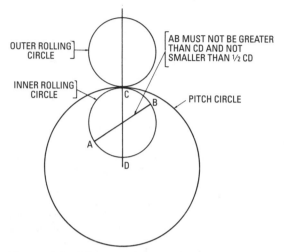

Figure 4-1 Diagram showing the generation of cycloid form of gear teeth.

If the diameter of the rolling circle is equal to the radius of the pitch circle, the hypocycloid will be a straight line that passes through the center of the pitch circle, and the flank of the tooth will be a radial. The same rolling circle must be used for interchangeable gears. In any other instance, the same rolling circles can be used for both gears, or their diameters can be used in proportion to their respective pitch circles.

A *cycloid* curve occurs when the generating circle rolls along a straight line. An *epicycloid* curve occurs when the generating circle (outer rolling circle) rolls along the outside of a circle. A *hypocycloid* curve occurs when the generating circle (inner rolling circle) rolls along the inside of a circle.

In comparing cycloid teeth with involute teeth, the involute gear teeth are stronger, run well with their centers at varying distances, and transmit uniform velocity. The chief objection to involute teeth is that they cause increased pressure on the bearings because of the obliquity of their action. However, modified tooth forms have been introduced on some involute gears.

Diametral and Circular Pitch Systems
Most of the cut gears in the United States are produced by the diametral pitch system. The circular pitch system is commonly used if the gear teeth are larger than *one* diametral pitch. Circular

pitch can also be applied to smaller gears if the required center-to-center distance cannot be obtained by a standard diametral pitch. The circular pitch system is also used on cast gears and on worm gearing, although these gears can also be designed for diametral pitch.

As mentioned, the diametral pitch system can be used to designate a series of standard gear tooth sizes (whole numbers), just as screw-thread pitches are standardized to designate the size of a thread. Since there must be a whole number of teeth on a gear, the increase in pitch diameter per tooth varies with the pitch. For example, the pitch diameter of a gear having 20 teeth of 4 diametral pitch is 5 inches. Therefore, the increase in diameter for each additional tooth is equal to ¼ inch per tooth. Similarly, for a diametral pitch of 2, the increase in diameter would be equal to ½ inch for each additional tooth. If a given center distance must be maintained and a standard diametral pitch cannot be used, it may be necessary to use gears based on the circular pitch system. Figure 4-2 shows gear teeth of different diametral pitch.

American Standard Spur Gear Tooth Forms

Four spur gear tooth forms are covered by the American Standards Association (ASA). Because the rack is the basis of a standard system of interchangeable gears, it is necessary only to give the proportions of the rack teeth in establishing a gear tooth standard.

American Standard 14.5° Involute Full-Depth Tooth

Figure 4-3 shows the rack of a standard 14.5° full-depth standard tooth form. This tooth form is very successful if the tooth numbers are large enough to avoid excessive undercutting of the teeth. Undercutting begins when the number of teeth is less than 32.

American Standard 20° Involute Full-Depth Tooth

The formulas for the 14.5° and 20° full-depth tooth standards are identical except for the radius at the base of the tooth. The pressure angle is the chief difference in the two standards. The larger pressure angle reduces undercutting, which begins when the number of teeth is less than 18 and may be excessive when the number of teeth is less than 14. The 20° teeth are wider at the base, which makes them stronger than the 14.5° teeth (see Figure 4-3).

The *American Standard 20° Involute Fine-Pitch Tooth* is a tooth form used for gears of 20 diametral pitch and finer. It is the same as the 20° full-depth tooth form except for a slight increase in whole depth. The additional depth and clearance are required because the

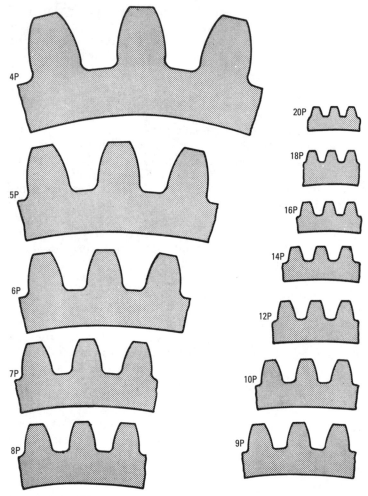

Figure 4-2 Gear teeth of different diametral pitch (full size).

(Courtesy Machinery's Handbook, Industrial Press)

wear on fine-pitch teeth is proportionally greater, the fillet radius is usually greater, and provision must be made for foreign material to accumulate at the bottom of the tooth spaces.

American Standard 14.5° Composite Tooth

This standard differs from the 14.5° involute full-depth system in regard to the form of the basic rack. The pressure angle and the

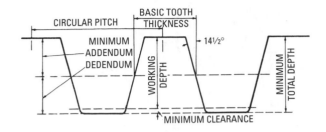

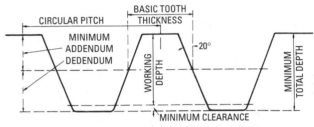

Figure 4-3 Basic rack of the 14.5° full-depth involute system (top) and the 20° full-depth involute system (bottom).
(Courtesy Machinery's Handbook, The Industrial Press)

various formulas for determining the tooth depth, addendum, and so on, are the same. The involute form of rack is modified by introducing a cycloidal curve below the pitch line and one above it to make the tooth symmetrical, as required for interchangeable gearing.

Since it is impractical to produce a rack with cycloidal curves in the shop, or a cutter of exact form, the approximate rack form (Figure 4-4) is used to meet the practical requirements. The curves are close approximations of the cycloidal curves on the theoretical rack.

The composite tooth form was developed originally for use with the form milling process. Gear teeth conforming to the composite standard usually are cut by form milling. This form of teeth can be produced readily on hobbing machines and other machines by making a hob or cutter of the basic rack form. The relieving tool can be made to the form of the basic rack if a hob is used.

American Standard 20° Involute Stub Tooth
The chief difference between the stub tooth and the full-depth tooth form is in regard to the tooth depth (see Figure 4-4). The shorter tooth, in combination with the 20° pressure angle, strengthens the stub form, and pinions with 12 and 13 teeth are undercut only slightly. The length of contact between mating gears is shortened, which tends to offset the increase in tooth strength and tends

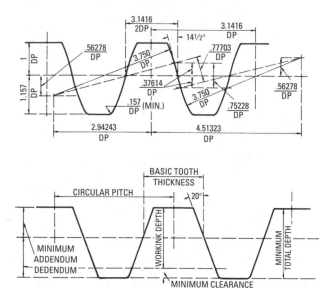

Figure 4-4 Basic rack of the 14.5° composite system (top) and the 20° stub involute system (bottom). *(Courtesy Machinery's Handbook, The Industrial Press)*

toward greater noise when the gears are running unless the tendency is offset by greater accuracy in cutting and mounting, This tendency to noise can be objectionable in certain classes of service, but it may be acceptable in other classes of service. For example, noise that is objectionable in automotive transmissions may not be a factor in gears used on some other classes of machinery.

The 20° stub tooth form is used extensively in automotive transmissions because the maximum power-transmitting capacity for a given pitch or material is essential and relatively small gears are required. However, very accurate gears are required, and the mountings are designed to minimize noise. Helical gear forms are also used because of their smooth, continuous action. The American Gear Manufacturers' Association recommends the American Standard 20° stub tooth system. These gears can be used interchangeably with other stub tooth systems, and only the amount of clearance is affected by the result of variations in tooth heights.

Fellows Stub Tooth

Two diametral pitches are used as a basis for the system of stub gear teeth introduced by the Fellows Gear Shaper Co. One diametral pitch (say, 8) is used as the basis for obtaining the dimension for the addendum and dedendum. Another diametral pitch (say, 6) is used to obtain the dimensions for the thickness of tooth, the number of teeth, and the pitch diameter. Thus, the teeth are designated as $6/8$ pitch. The numerator of the fraction indicates the pitch used to obtain the tooth thickness and the number of teeth, and the denominator indicates the pitch used to determine the depth of the tooth. The clearance is greater than in the ordinary gear tooth system. The clearance angle is 20°.

Nuttall Stub Tooth
In this system, tooth dimensions are based directly on the circular pitch. The addendum is equal to 0.250 times the circular pitch, and the dedendum is equal to 0.300 times the circular pitch. The pressure angle for this system is 20°.

Gear-Cutting Operations
Formerly, all gears were cut on milling machines, a cutter being used to produce the correct tooth shape. Then, the milling machine was replaced with automatic gear cutting machines, especially for production work.

In gear cutting on the milling machine, the first problem for the machinist is to select the proper cutter because the shape of the tooth changes with the number of gear teeth. For example, the exact shape of the tooth of a 179-tooth gear is slightly different from that of a gear with 180 teeth. Although the difference is extremely small in this instance, it would be a much greater difference in gears of 20 and 21 teeth. The difference in shape is more marked in gears that have a relatively smaller number of teeth. For most practical purposes, these variations in shape can be ignored.

Cutting Spur Gears
Spur gears have straight teeth, cut parallel with the axis of rotation. Standard involute gear cutters of the arbor-mounted type are usually used to mill spur gears (Figure 4-5).

Selection of Cutter
Standard sets of cutters are made by the manufacturers. For example, the Cincinnati Milacron Co. and others make eight cutters as follows:

- No. 1—135 teeth to a rack
- No. 2—55–134 teeth
- No. 3—35–54 teeth

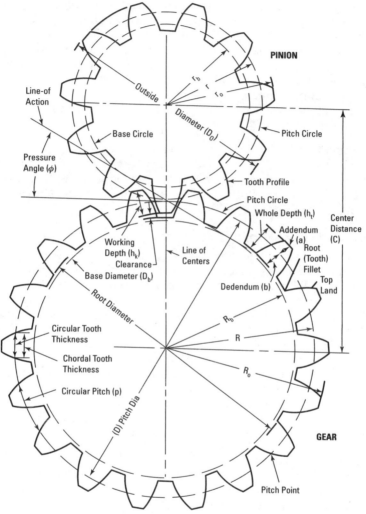

Figure 4-5 Basic spur gear geometry.

- No. 4—26–34 teeth
- No. 5—21–25 teeth
- No. 6—17–20 teeth
- No. 7—14–16 teeth
- No. 8—12 and 13 teeth

For a finer division of the number of teeth that can be cut, the Brown and Sharpe Co. manufactures cutters in half-numbers as follows:

- No. 1½—80–134 teeth
- No. 2½—42–54 teeth
- No. 3½—30–34 teeth
- No. 4½—23–25 teeth
- No. 5½—19 and 20 teeth
- No. 6½—15 and 16 teeth
- No. 7½—13 teeth

To select the proper cutter, it is necessary to know the number of teeth and either the diametral or the circular pitch. For example, to cut a gear with 48 teeth, first note that a No. 3 cutter can be used to cut gears with 35 to 54 teeth. The shape of the tooth that is produced will not be entirely accurate, but it will be sufficiently accurate for uses in which high speeds and smoothness in running are not essential. If the gears must be more accurate, a special cutter made to the correct shape for a given number of teeth must be used. Cutter manufacturers can furnish these special cutters.

It is not advisable to install an automatic gear-cutting machine in a shop unless a large amount of gear-cutting work must be performed. A milling machine can cut gears as rapidly as an automatic gear cutter, and it requires no more time to set up.

The chief advantage of the automatic gear-cutting machine (compared with the milling machine) lies in the fact that it can automatically index the gear and return the cutter at a rapid rate. The gear cutter operates automatically after the cut has been started. The milling machine requires the attention of the operator for indexing and advancing the work and for throwing in the feed. Even shops that have the automatic gear cutters frequently use the milling machine for odd jobs of gear cutting because it lends itself to rapid setup with no special preparation for indexing.

Setup

The gear bank is placed on a mandrel held between centers on a universal dividing head (Figure 4-6). The dividing head is located on the table of a universal milling machine.

In gear cutting, it is especially important that there is no backlash in the indexing mechanism. The index pin should be brought around

Figure 4-6   Using the high-number index plate to mill a spur gear.
(Courtesy Cincinnati Milacron Co.)

in the direction that the indexing will be done, which is preferably a clockwise direction, and permitted to drop into one of the holes.

Then, set the sector for proper spacing, tighten the spindle clamp at the rear of the dividing head, and start the machine. The work should be elevated carefully until the revolving cutter barely touches the work. Then, set the elevating dial at zero, run the table to the right-hand side until the cutter is cleared, and elevate for the proper depth, as indicated on the dial.

Disengage the elevating crank to reduce the possibility of disturbing the setting. Then, the setup is complete for beginning the milling operation. In adjusting the depth, make certain that backlash is removed before the final setting is made.

Measurement

It is customary to take a trial cut on each side of one tooth just enough to mill the full outline of the tooth. The machine is then stopped and the tooth thickness measured with a vernier gear-tooth caliper (see Figure 4-7).

The gear-tooth vernier caliper measures chordal thickness (or thickness at the pitch line of a gear tooth) to one-thousandth of

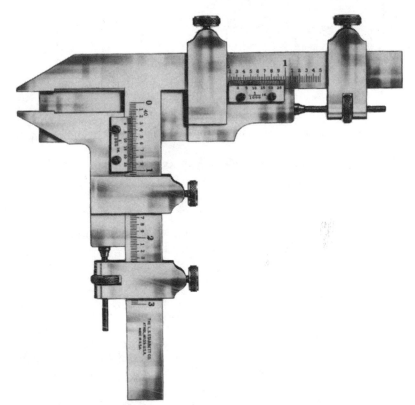

Figure 4-7 Vernier gear-tooth caliper. *(Courtesy L.S. Starrett Co.)*

an inch. Its construction combines in one tool the function of both vernier depth gage and vernier caliper. The vertical slide is set to depth by means of its vernier plate fine-adjusting nut so that when it rests on top of the gear tooth, the caliper jaws will be correctly positioned to measure across the pitch line of the gear tooth. The horizontal slide is then used to obtain the chordal thickness of the gear tooth by means of its vernier slide fine-adjusting nut.

The procedure for reading these gages is exactly the same as for vernier calipers. It is necessary to determine the correct chordal thickness and chordal addendum (or *corrected addendum*). These measurements are illustrated in Figure 4-8.

The chordal, or straight-line, thickness of a standard gear tooth can be found by the following formula, where T_c = chordal thickness,

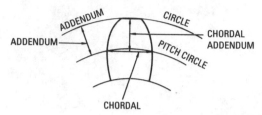

Figure 4-8 Measuring the chordal thickness and chordal addendum.

D = pitch diameter, and N = number of teeth:

$$T_c = D \sin \frac{90°}{N}$$

As an example, consider a gear that has 15 teeth and a pitch diameter of 5 inches. What is the chordal thickness?

$$T_c = 5 \sin \frac{90°}{15} = 5 \sin 6°$$

$$T_c = 5 \times 0.10453 = 0.5226 \text{ inch}$$

In measuring the chordal thickness, the vertical scale of the gear-tooth caliper is set to the chordal or corrected addendum to locate the caliper jaws at the pitch line. The following formula is used in determining the chordal addendum, where a_c = chordal addendum, a = addendum, and T = circular thickness of tooth at pitch diameter D:

$$a_c = a + \frac{T^2}{4D}$$

As an example, suppose a gear has a 1.75-inch pitch diameter, a tooth thickness of 0.2176, and an addendum of 0.1542. What is the chordal addendum?

$$a_c = 0.1542 + \frac{0.2176^2}{4 + 1.75}$$

$$a_c = 0.1610 \text{ inch}$$

Chordal thicknesses and chordal addenda can be secured by using Table 4-1. Figure 4-9 shows the correct way to hold a vernier gear-tooth caliper.

Table 4-1   Chordal Thickness of Gear Teeth Bases of 1 Diametral Pitch

No. of Teeth	t''	s''	No. of Teeth	t''	s''	No. of Teeth	t''	s''
6	1.5529	1.1022	40	1.5704	1.0154	74	1.5707	1.0084
7	1.5568	1.0873	41	1.5704	1.0150	75	1.5707	1.0083
8	1.5607	1.0769	42	1.5704	1.0147	76	1.5707	1.0081
9	1.5628	1.0684	43	1.5705	1.0143	77	1.5707	1.0080
10	1.5643	1.0616	44	1.5705	1.0140	78	1.5707	1.0079
11	1.5654	1.0559	45	1.5705	1.0137	79	1.5707	1.0078
12	1.5663	1.0514	46	1.5705	1.0134	80	1.5707	1.0077
13	1.5670	1.0474	47	1.5705	1.0131	81	1.5707	1.0076
14	1.5675	1.0440	48	1.5705	1.0129	82	1.5707	1.0075
15	1.5679	1.0411	49	1.5705	1.0126	83	1.5707	1.0074
16	1.5683	1.0385	50	1.5705	1.0123	84	1.5707	1.0074
17	1.5686	1.0362	51	1.5706	1.0121	85	1.5707	1.0073
18	1.5688	1.0342	52	1.5706	1.0119	86	1.5707	1.0072
19	1.5690	1.0324	53	1.5706	1.0117	87	1.5707	1.0071
20	1.5692	1.0308	54	1.5706	1.0114	88	1.5707	1.0070
21	1.5694	1.0294	55	1.5706	1.0112	89	1.5707	1.0069
22	1.5695	1.0281	56	1.5706	1.0110	90	1.5707	1.0068
23	1.5696	1.0268	57	1.5706	1.0108	91	1.5707	1.0068
24	1.5697	1.0257	58	1.5706	1.0106	92	1.5707	1.0067
25	1.5698	1.0247	59	1.5706	1.0105	93	1.5707	1.0067
26	1.5698	1.0237	60	1.5706	1.0102	94	1.5707	1.0066
27	1.5699	1.0228	61	1.5706	1.0101	95	1.5707	1.0065
28	1.5700	1.0220	62	1.5706	1.0100	96	1.5707	1.0064
29	1.5700	1.0213	63	1.5706	1.0098	97	1.5707	1.0064
30	1.5701	1.0208	64	1.5706	1.0097	98	1.5707	1.0063
31	1.5701	1.0199	65	1.5706	1.0095	99	1.5707	1.0062
32	1.5702	1.0193	66	1.5706	1.0094	100	1.5707	1.0061
33	1.5702	1.0187	67	1.5706	1.0092	101	1.5707	1.0061
34	1.5702	1.0181	68	1.5706	1.0091	102	1.5707	1.0060
35	1.5702	1.0176	69	1.5707	1.0090	103	1.5707	1.0060
36	1.5703	1.0171	70	1.5707	1.0088	104	1.5707	1.0059
37	1.5703	1.0167	71	1.5707	1.0087	105	1.5707	1.0059
38	1.5703	1.0162	72	1.5707	1.0086	106	1.5707	1.0058
39	1.5704	1.0158	73	1.5707	1.0085	107	1.5707	1.0058

(continued)

Table 4-1 (continued)

No. of Teeth	t''	s''	No. of Teeth	t''	s''	No. of Teeth	t''	s''
108	1.5707	1.0057	119	1.5707	1.0052	130	1.5707	1.0047
109	1.5707	1.0057	120	1.5707	1.0052	131	1.5708	1.0047
110	1.5707	1.0056	121	1.5707	1.0051	132	1.5708	1.0047
111	1.5707	1.0056	122	1.5707	1.0051	133	1.5708	1.0047
112	1.5707	1.0055	123	1.5707	1.0050	134	1.5708	1.0046
113	1.5707	1.0055	124	1.5707	1.0050	135	1.5708	1.0046
114	1.5707	1.0054	125	1.5707	1.0049	150	1.5708	1.0045
115	1.5707	1.0054	126	1.5707	1.0049	250	1.5708	1.0025
116	1.5707	1.0053	127	1.5707	1.0049	Rack	1.5708	1.0000
117	1.5707	1.0053	128	1.5707	1.0048			
118	1.5707	1.0053	129	1.5707	1.0048			

S = module or addendum, or distance from top to pitch line of tooth

s'' = corrected $S = H + S$

t'' = chordal thickness of tooth

H = height of arc

When using gear-tooth vernier caliper to measure coarse pitch gear teeth, the chordal thickness t'' must be known, since t'' is less than the regular thickness AB measured on the pitch line. In referring to the table above, note that height of arc H has been added to the addendum S, the corrected figures to use being found in column s''.

For any other pitch, divide figures in table by the required pitch.

(Courtesy L. S. Starrett Co.)

Cutting Bevel Gears

Bevel gears are conical gears (cone-shaped) and are used to connect shafts that have intersecting axes. Hypoid gears are similar to bevel gears in their general form, but they operate on axes that are offset. Most bevel gears can be classified as either the *straight-tooth* type or the *curved-tooth* type. Spiral bevels, Zerol bevels, and hypoid gears are all classified as curved-tooth gears. Table 4-2 shows the number of formed cutters used to mill teeth in mating bevel gears and pinions with shafts at right angles. The number of cutters for the gear is provided first, followed by the number for the pinion.

Straight bevel gears are the most commonly used type of all the bevel gears. The teeth are straight, but their sides are tapered, so that they would intersect the axis at a common point called the *pitch cone apex* if they were extended inwardly. In most straight bevel gears, the face cone elements are made parallel to the root

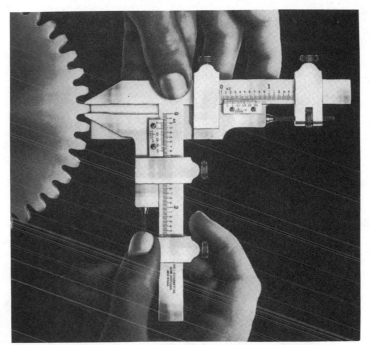

Figure 4-9   Correct use of the vernier gear-tooth caliper.
(Courtesy L.S. Starrett Co.)

cone elements of the mating gear to obtain uniform clearance along the length of the teeth. Therefore, the face cone elements intersect the axis at a point inside the pitch cone. The calculations for the straight bevel gears are the easiest to perform. Straight bevel gears are also the most economical of the types of bevel gears to produce.

Straight bevel gear teeth can be generated either for full contact or for localized contact. Gear teeth developed for localized contact are slightly convex in a lengthwise direction so that some adjustment of the gears during assembly is possible, and small displacements caused by load deflections can take place without an undesirable load concentration on the ends of the teeth. The slight lengthwise rounding of the sides of the teeth does not have to be computed in the tooth design, but it is taken care of automatically in the cutting operations performed on the newer bevel gear generators. Figure 4-10 shows a set of straight bevel gears (a gear and a pinion) being power tested. These have been generated for localized contact, and the dark spots on each tooth indicate the contact area.

Table 4-2 Bevel Gearing

								Number of Teeth in Pinion										
Number of Teeth in Gear	12	13	14	15	16	17	18	19	20	21	22	23	24	25	26	27	28
12	7-7																
13	6-7	6-6															
14	5-7	6-6	6-6														
15	5-7	5-6	5-6	5-5													
16	4-7	5-7	5-6	5-6	5-5												
17	4-7	4-7	4-6	5-6	5-5	5-5											
18	4-7	4-7	4-6	4-6	4-5	4-5	5-5										
19	3-7	4-7	4-6	4-6	4-6	4-5	4-5	4-4									
20	3-7	3-7	4-6	4-6	4-6	4-5	4-5	4-4	4-4								
21	3-8	3-7	3-7	3-6	4-6	4-5	4-5	4-5	4-4	4-4							
22	3-8	3-7	3-7	3-6	3-6	4-5	4-5	4-5	4-4	4-4	4-4						
23	3-8	3-7	3-7	3-6	3-6	3-5	3-5	3-5	3-4	4-4	4-4	4-4					
24	3-8	3-7	3-7	3-6	3-6	3-6	3-5	3-5	3-4	3-4	3-4	4-4	4-4				
25	2-8	2-7	3-7	3-6	3-6	3-6	3-5	3-5	3-5	3-4	3-4	3-4	4-4	3-3			
26	2-8	2-7	3-7	3-6	3-6	3-6	3-5	3-5	3-5	3-4	3-4	3-4	3-4	3-3	3-3		
27	2-8	2-7	2-7	2-6	3-6	3-6	3-5	3-5	3-5	3-4	3-4	3-4	3-4	3-4	3-3	3-3	
28	2-8	2-7	2-7	2-6	2-6	3-6	3-5	3-5	3-5	3-4	3-4	3-4	3-4	3-4	3-3	3-3	3-3
29	2-8	2-7	2-7	2-7	2-6	2-6	3-5	3-5	3-5	3-4	3-4	3-4	3-4	3-4	3-3	3-3	3-3
30	2-8	2-7	2-7	2-7	2-6	2-5	2-5	3-5	3-5	3-5	3-4	3-4	3-4	3-4	3-4	3-3	3-3

#																	
31	2-8	2-7	2-7	2-7	2-7	2-6	2-6	2-5	2-5	2-5	3-4	3-4	3-4	3-4	3-4	3-3	3-3
32	2-8	2-7	2-7	2-7	2-7	2-6	2-6	2-5	2-5	2-5	2-4	2-4	3-4	3-4	3-4	3-3	3-3
33	2-8	2-8	2-7	2-7	2-7	2-6	2-6	2-5	2-5	2-5	2-4	2-4	2-4	3-4	3-4	3-4	3-3
34	2-8	2-8	2-7	2-7	2-7	2-6	2-6	2-5	2-5	2-5	2-4	2-4	2-4	2-4	2-4	3-4	3-3
35	2-8	2-8	2-7	2-7	2-7	2-6	2-6	2-5	2-5	2-5	2-5	2-4	2-4	2-4	2-4	3-3	2-3
36	2-8	2-8	2-7	2-7	2-7	2-6	2-6	2-5	2-5	2-5	2-5	2-4	2-4	2-4	2-4	2-4	2-3
37	2-8	2-8	2-7	2-7	2-7	2-6	2-6	2-5	2-5	2-5	2-5	2-4	2-4	2-4	2-4	2-4	2-3
38	2-8	2-8	2-7	2-7	2-7	2-6	2-6	2-5	2-5	2-5	2-5	2-4	2-4	2-4	2-4	2-4	2-4
39	2-8	2-8	2-7	2-7	2-7	2-6	2-6	2-5	2-5	2-5	2-5	2-4	2-4	2-4	2-4	2-4	2-4
40	1-8	2-8	2-7	2-7	2-7	2-6	2-6	2-5	2-5	2-5	2-5	2-5	2-4	2-4	2-4	2-4	2-4
41	1-8	1-8	2-7	2-7	2-7	2-6	2-6	2-6	2-5	2-5	2-5	2-5	2-4	2-4	2-4	2-4	2-4
42	1-8	1-8	1-7	1-7	2-7	2-6	2-6	2-6	2-5	2-5	2-5	2-5	2-4	2-4	2-4	2-4	2-4
43	1-8	1-8	1-7	1-7	2-7	2-6	2-6	2-6	2-5	2-5	2-5	2-5	2-4	2-4	2-4	2-4	2-4
44	1-8	1-8	1-7	1-7	1-6	2-6	2-6	2-6	2-5	2-5	2-5	2-5	2-4	2-4	2-4	2-4	2-4
45	1-8	1-8	1-7	1-7	1-7	2-6	2-6	2-6	2-5	2-5	2-5	2-5	2-4	2-4	2-4	2-4	2-4
46	1-8	1-8	1-7	1-7	1-7	2-6	2-6	2-6	2-5	2-5	2-5	2-5	2-4	2-4	2-4	2-4	2-4
47	1-8	1-8	1-7	1-7	1-7	1-6	2-6	2-6	2-5	2-5	2-5	2-5	2-4	2-4	2-4	2-4	2-4
48	1-8	1-8	1-7	1-7	1-7	1-6	2-6	2-6	2-5	2-5	2-5	2-5	2-4	2-4	2-4	2-4	2-4
49	1-8	1-8	1-7	1-7	1-7	1-6	1-6	1-6	2-5	2-5	2-5	2-5	2-4	2-4	2-4	2-4	2-4
50	1-8	1-8	1-7	1-7	1-7	1-6	1-6	1-6	2-5	2-5	2-5	2-5	2-4	2-4	2-4	2-4	2-4
51	1-8	1-8	1-7	1-7	1-7	1-6	1-6	1-6	1-5	1-5	2-5	2-5	2-4	2-4	2-4	2-4	2-4
52	1-8	1-8	1-7	1-7	1-7	1-6	1-6	1-6	1-5	1-5	2-5	2-5	2-4	2-4	2-4	2-4	2-4
53	1-8	1-8	1-7	1-7	1-7	1-6	1-6	1-6	1-5	1-5	2-5	2-5	2-4	2-4	2-4	2-4	2-4
54	1-8	1-8	1-7	1-7	1-7	1-6	1-6	1-6	1-5	1-5	1-5	1-5	2-4	2-4	2-4	2-4	2-4
55	1-8	1-8	1-7	1-7	1-7	1-6	1-6	1-6	1-5	1-5	1-5	1-5	1-4	2-4	2-4	2-4	2-4

Table 4-2 (continued)

| | | | | | | | | Number of Teeth in Pinion | | | | | | | | |
12	13	14	15	16	17	18	19	20	21	22	23	24	25	26	27	28
56	1–8	1–8	1–8	1–7	1–7	1–6	1–6	1–6	1–6	1–5	1–5	1–5	1–5	1–4	1–4	2–4	2–4
57	1–8	1–8	1–8	1–7	1–7	1–6	1–6	1–6	1–6	1–5	1–5	1–5	1–5	1–4	1–4	1–4	2–4
58	1–8	1–8	1–8	1–7	1–7	1–6	1–6	1–6	1–6	1–5	1–5	1–5	1–5	1–4	1–4	2–4	2–4
59	1–8	1–8	1–8	1–7	1–7	1–6	1–6	1–6	1–6	1–5	1–5	1–5	1–5	1–4	1–4	2–4	2–4
60	1–8	1–8	1–8	1–7	1–7	1–6	1–6	1–6	1–6	1–5	1–5	1–5	1–5	1–4	1–4	2–4	2–4
61	1–8	1–8	1–8	1–7	1–7	1–6	1–6	1–6	1–6	1–5	1–5	1–5	1–5	1–4	1–4	2–4	2–4
62	1–8	1–8	1–8	1–7	1–7	1–6	1–6	1–6	1–6	1–5	1–5	1–5	1–5	1–4	1–4	2–4	2–4
63	1–8	1–8	1–8	1–7	1–7	1–6	1–6	1–6	1–6	1–5	1–5	1–5	1–5	1–4	1–4	2–4	2–4
64	1–8	1–8	1–8	1–7	1–7	1–6	1–6	1–6	1–6	1–5	1–5	1–5	1–5	1–4	1–4	2–4	2–4
65	1–8	1–8	1–8	1–7	1–7	1–6	1–6	1–6	1–6	1–5	1–5	1–5	1–5	1–4	1–4	2–4	2–4
66	1–8	1–8	1–8	1–7	1–7	1–6	1–6	1–6	1–6	1–5	1–5	1–5	1–5	1–4	1–4	2–4	2–4
67	1–8	1–8	1–8	1–7	1–7	1–6	1–6	1–6	1–6	1–5	1–5	1–5	1–5	1–4	1–4	2–4	2–4
68	1–8	1–8	1–8	1–7	1–7	1–6	1–6	1–6	1–6	1–5	1–5	1–5	1–5	1–4	1–4	2–4	2–4
69	1–8	1–8	1–8	1–7	1–7	1–6	1–6	1–6	1–6	1–5	1–5	1–5	1–5	1–4	1–4	2–4	2–4
70	1–8	1–8	1–8	1–7	1–7	1–6	1–6	1–6	1–6	1–5	1–5	1–5	1–5	1–4	1–4	2–4	2–4
71	1–8	1–8	1–8	1–7	1–7	1–6	1–6	1–6	1–6	1–5	1–5	1–5	1–5	1–4	1–4	2–4	2–4
72	1–8	1–8	1–8	1–7	1–7	1–6	1–6	1–6	1–6	1–5	1–5	1–5	1–5	1–4	1–4	2–4	2–4
73	1–8	1–8	1–8	1–7	1–7	1–6	1–6	1–6	1–6	1–5	1–5	1–5	1–5	1–4	1–4	2–4	2–4
74	1–8	1–8	1–8	1–7	1–7	1–6	1–6	1–6	1–6	1–5	1–5	1–5	1–5	1–4	1–4	2–4	2–4
75	1–8	1–8	1–8	1–7	1–7	1–6	1–6	1–6	1–6	1–5	1–5	1–5	1–5	1–4	1–4	2–4	2–4
76	1–8	1–8	1–8	1–7	1–7	1–6	1–6	1–6	1–6	1–5	1–5	1–5	1–5	1–4	1–4	2–4	2–4

77	1–8	1–8	1–7	1–7	1–7	1–6	1–6	1–6	1–5	1–5	1–5	1–5	1–4	1–4	2–4	2–4	2–4
78	1–8	1–8	1–7	1–7	1–7	1–6	1–6	1–6	1–5	1–5	1–5	1–5	1–4	1–4	2–4	2–4	2–4
79	1–8	1–8	1–7	1–7	1–7	1–6	1–6	1–6	1–5	1–5	1–5	1–5	1–4	1–4	2–4	2–4	2–4
80	1–8	1–8	1–7	1–7	1–7	1–6	1–6	1–6	1–5	1–5	1–5	1–5	1–4	1–4	2–4	2–4	2–4
81	1–8	1–8	1–7	1–7	1–7	1–6	1–6	1–6	1–5	1–5	1–5	1–5	1–4	1–4	2–4	2–4	2–4
82	1–8	1–8	1–7	1–7	1–7	1–6	1–6	1–6	1–5	1–5	1–5	1–5	1–4	1–4	2–4	2–4	2–4
83	1–8	1–8	1–7	1–7	1–7	1–6	1–6	1–6	1–5	1–5	1–5	1–5	1–4	1–4	2–4	2–4	2–4
84	1–8	1–8	1–7	1–7	1–7	1–6	1–6	1–6	1–5	1–5	1–5	1–5	1–4	1–4	2–4	2–4	2–4
85	1–8	1–8	1–7	1–7	1–7	1–6	1–6	1–6	1–5	1–5	1–5	1–5	1–4	1–4	2–4	2–4	2–4
86	1–8	1–8	1–7	1–7	1–7	1–6	1–5	1–6	1–5	1–5	1–5	1–5	1–4	1–4	2–4	2–4	2–4
87	1–8	1–8	1–7	1–7	1–7	1–6	1–6	1–6	1–5	1–5	1–5	1–5	1–4	1–4	2–4	2–4	2–4
88	1–8	1–8	1–7	1–7	1–7	1–6	1–6	1–6	1–5	1–5	1–5	1–5	1–4	1–4	2–4	2–4	2–4
89	1–8	1–8	1–7	1–7	1–7	1–6	1–6	1–6	1–5	1–5	1–5	1–5	1–4	1–4	2–4	2–4	2–4
90	1–8	1–8	1–7	1–7	1–7	1–6	1–6	1–6	1–5	1–5	1–5	1–5	1–4	1–4	2–4	2–4	2–4
91	1–8	1–8	1–7	1–7	1–7	1–6	1–6	1–6	1–5	1–5	1–5	1–5	1–4	1–4	2–4	2–4	2–4
92	1–8	1–8	1–7	1–7	1–7	1–6	1–6	1–6	1–5	1–5	1–5	1–5	1–4	1–4	2–4	2–4	2–4
93	1–8	1–8	1–7	1–7	1–7	1–6	1–6	1–6	1–5	1–5	1–5	1–5	1–4	1–4	2–4	2–4	2–4
94	1–8	1–8	1–7	1–7	1–7	1–6	1–6	1–6	1–5	1–5	1–5	1–5	1–4	1–4	2–4	2–4	2–4
95	1–8	1–8	1–7	1–7	1–7	1–6	1–6	1–6	1–5	1–5	1–5	1–5	1–4	1–4	2–4	2–4	2–4
96	1–8	1–8	1–7	1–7	1–7	1–6	1–6	1–6	1–5	1–5	1–5	1–5	1–4	1–4	2–4	2–4	2–4
97	1–8	1–8	1–7	1–7	1–7	1–6	1–6	1–6	1–5	1–5	1–5	1–5	1–4	1–4	2–4	2–4	2–4
98	1–8	1–8	1–7	1–7	1–7	1–6	1–6	1–6	1–5	1–5	1–5	1–5	1–4	1–4	2–4	2–4	2–4
99	1–8	1–8	1–7	1–7	1–7	1–6	1–6	1–6	1–5	1–5	1–5	1–5	1–4	1–4	2–4	2–4	2–4
100	1–8	1–8	1–7	1–7	1–7	1–6	1–6	1–6	1–5	1–5	1–5	1–5	1–4	1–4	2–4	2–4	2–4

Figure 4-10   Straight bevel gears under test. *(Courtesy Illinois Gear)*

Zerol bevel gears have curved teeth, but they lie in the same general direction as the teeth of straight bevel gears. These bevel gears can be considered to be spiral bevel gears with zero spiral angle, and they are manufactured on the same machines as spiral bevel gears. In Zerol bevel gears, the face cone elements do not pass through the pitch cone apex, but they are approximately parallel to the root cone elements of the mating gear to provide uniform tooth clearance. Zerol bevel gears can be used when highly accurate (produced by grinding) hardened bevel gears are required.

Spiral bevel gears have curved oblique teeth. Contact begins gradually and continues smoothly from end to end. They mesh with a rolling contact that is similar to straight bevel gears. As a result of their overlapping tooth action, spiral gears transmit motion more smoothly than either straight bevel or Zerol bevel gears, thereby reducing noise and vibration, which becomes especially noticeable at high speeds. Localized tooth contact promotes smooth, quiet-running spiral bevel gears and permits some deflection in mounting without concentrating the load near either end of the tooth. One of the advantages of spiral bevel gears is the complete control of localized tooth contact. The amount of surface over which tooth contact takes place can be changed to suit the specific requirements of each job by making a slight change in the radii of the curvature of the matching tooth surfaces.

Because their tooth surfaces can be ground, spiral bevel gears are advantageous in applications that require hardened gears of high accuracy. The bottoms of the tooth spaces and the tooth profiles can be ground simultaneously. This results in a smooth blending of the tooth profile, the tooth fillet, and the bottom of the tooth space. This is important from the standpoint of strength because it eliminates cutter marks and other surface irregularities that often result in stress concentrations. In some cases, the gears are lapped to improve the accuracy and finish. Figure 4-11 shows the lapping of intermediate-size spiral bevel gears.

Figure 4-11   Lapping spiral bevel gears. *(Courtesy Illinois Gear)*

Hypoid gears resemble spiral bevel gears in general appearance, except that the axis of the pinion is offset in relation to the gear axis (Figure 4-12). If the offset is sufficient, the shafts may pass one another, thus permitting a compact straddle mounting on the gear and pinion. Because a spiral bevel pinion has equal pressure angles

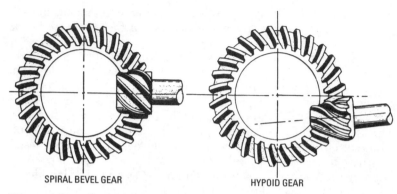

SPIRAL BEVEL GEAR   HYPOID GEAR

Figure 4-12 Comparison of the spiral bevel gears (left) and hypoid gears (right), showing offset axis in the hypoid gears.

and symmetrical profile curvatures on both sides of the teeth, a hypoid pinion that is properly conjugated to a mating gear having equal pressure angles on both sides of the teeth must have nonsymmetrical profile curvatures for proper tooth action. To obtain equal arcs of motion for both sides of the teeth, it is necessary to use unequal pressure angles on hypoid pinions.

The hypoid gears are usually designed so that the pinion has a larger spiral angle than the gear. The advantage of such design is that the pinion diameter is increased and is stronger than a corresponding spiral bevel pinion. This increment of diameter permits the use of comparatively high ratios without the pinion becoming too small to allow a bore or shank of adequate size.

The sliding action along the lengthwise direction of their teeth, in hypoid gears, is a function of the difference in the spiral angles on the gear and pinion. This sliding effect makes the gears run more smoothly than spiral bevel gears. Hypoid gears can be ground on the same machines that are used to grind Zerol bevel gears and spiral bevel gears. Hypoid gears are frequently lapped to improve their accuracy and finish.

As mentioned previously, bevel gears are cone-shaped. In form, they resemble the frustums of cones. The axes of bevel gears intersect at a common apex to form an obtuse, acute, or right angle. Figure 4-13 shows the evolution of bevel gears. To draw the extended cones M and N, let the line YY represent the axis of one of the cones. Through any point on the axis YY, draw line AB at a right angle to line YY, making line AB equal to line BC. Thus, line AC is the base of the cone M.

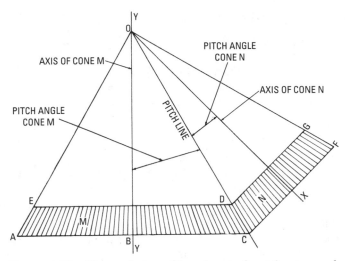

Figure 4-13   The evolution of bevel gears from frustums of cones.

Using any point O on line Y as the apex (depending on the slant angle desired), draw lines OA and OC. Line OC represents the pitch line (corresponding to the pitch circle of a spur gear). The angle BOC is the pitch cone angle, and the distance OC is the pitch cone radius. Lay off the distance CD on the pitch line equal to the desired slant length of the frustum, and draw line DE parallel with the baseline AC. The frustum of the first cone M is ACDF. The *frustum* is the part of a conical-shaped solid formed by cutting off the top by a plane parallel to the base.

For the second cone N, draw the axis OS through point O at an angle COX equal to the pitch angle of the cone N. Also, draw the line OF at an equal angle with the axis OX. From the points C and D, draw the lines CF and DG at 90° to the axis OX. Then, the frustum CFGD is the frustum of the second cone N. These two frustums M and N are in contact along the common line DC, which is the common pitch line (as in the two cylinders rolling in contact). Therefore, the two frustums can be considered to be two bevel gears that have an infinite number of teeth.

In actual operation, the frustum N revolves about its axis OX, and the frustum M revolves about its axis OY. If power is applied to the frustum N, it will drive the frustum M by frictional contact along the common pitch line CD. The relative speed of each is inversely proportional to the diameters, or to the number of teeth in the gears. The pitch line and the pitch angle are basic lines in the laying out of bevel gears.

Following are important definitions relative to bevel gears:

- *Pitch line*—A straight line that passes through the apex of the cone and lies in the slant surface; an element of the cone.
- *Pitch angle*—The angle between the pitch line and the axis of the cone.
- *Pitch cone radius*—The length of the pitch line from the apex of the cone to its base.
- *Addendum*—The distance the tooth extends outside the pitch line at the outer edge.
- *Dedendum*—The depth of the tooth below the pitch line at the outer edge.
- *Addendum angle*—The angle at the apex between the pitch line and the top line of the tooth.
- *Dedendum angle*—The angle at the apex between the pitch line and the baseline of the tooth.
- *Root angle*—The angle at the apex between the baseline of the tooth and the cone axis.
- *Force angle*—The pitch cone angle plus the addendum angle.
- *Pitch diameter*—The diameter of the vase of the cone.
- *Angular addendum*—The distance along the outer edge of the tooth from the cone axis less one-half the pitch diameter.
- *Outside diameter*—The pitch diameter plus two times the angular addendum.
- *Cutting angle*—The angle between the tooth baseline and the axis of the cone.

Figure 4-14 illustrates these definitions. First, draw the axis YY for the outline of the bevel gear, and generate the cone OAC from the given data. This determines the pitch line on either side OA or OC. Using the addendum and dedendum angles from given data, draw the top line and the baseline on both sides.

Through the points A and C, draw lines perpendicular to the pitch line (that is, perpendicular to lines OA and OC). The intersection of these lines with the pitch cone radius determines the outer edge of the tooth.

The inner edge of the tooth is determined from data giving the length of the tooth. The cup depth and cup clearance are laid out from the given data.

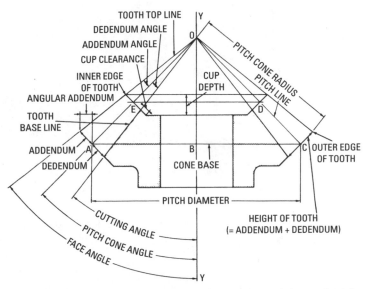

Figure 4-14   A method of obtaining the outline of the cycloidal tooth by generating the cone (OAC) and the frustum (ACDE) of the cone, for laying out a bevel gear.

Selection of Cutter

As in cutting spur gears, the first problem encountered is selection of the proper cutter for cutting the bevel gears (Figure 4-15). The length of the teeth or the face on bevel gears is usually not more than one-third the apex distance Ab, and the cutters that are normally carried in stock can be used. The difference between formed cutters used for milling spur gears and those used for bevel gears is that bevel gear cutters are thinner since they must pass through the narrow tooth space at the small end of the bevel gear. Otherwise, the shape of the cutter (and, hence, the cutter number) is the same. If the face is longer than one-third the apex distance, special thin cutters must be used.

In selecting the cutter, measure the back cone radius ab for the gear, or bc for the pinion. This is equal to the radius of a spur gear in which the number of teeth would determine the cutter to use. Twice the back cone radius ($2ab$) multiplied by the diametral pitch gives the number of teeth for which the cutter should be selected to cut the gear. The correct number of the cutter can be determined from the list of cutters.

As an example, determine the cutter to use for cutting a bevel gear having a back cone radius ab of 4 inches and a diametral pitch

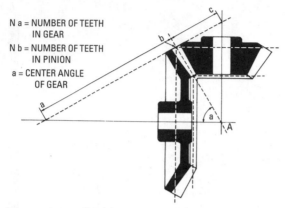

N a = NUMBER OF TEETH
    IN GEAR
N b = NUMBER OF TEETH
    IN PINION
a = CENTER ANGLE
    OF GEAR

Figure 4-15 Diagram showing the problem encountered in selection of a cutter for cutting bevel gears.

of 8. The formula is $(2ab \times \text{diametral pitch}) = \text{number of teeth}$. Substituting in the formula,

$$\text{Number of teeth} = 2\,(4) \times 8$$
$$= 8 \times 8$$
$$= 64$$

By referring to a list of standard cutters, we can select cutter No. 2 because the number of teeth (64) comes within the range (55–134) of the standard No. 2 cutter.

The number of teeth for which the cutter should be selected can also be found by the following formula:

$$\tan a = \frac{Na}{Nb}$$

To select the cutter for the number of teeth in the gear,

$$\text{Number of teeth} = \frac{Na}{\cos a}$$

To select the cutter for the number of teeth in the pinion,

$$\text{Number of teeth} = \frac{Nb}{\sin a}$$

If the gears are the same size (usually called *miter gears*), a single cutter can be used for both gears. Figure 4-16 shows the power

Figure 4-16 Power testing spiral miter gears. *(Courtesy Illinois Gear)*

testing of large spiral miter gears. Notice that the tooth size is being measured with a vernier gear-tooth caliper. Two cutters may be required, if one gear is larger than the other.

Tables can be used to select the correct formed cutter for milling bevel gears. Table 4-2 gives the number of cutters to use for milling various numbers of teeth in the gear and pinion. This table applies only to bevel gears with axes at right angles.

Offset of the Cutter
Because thickness of the cutter cannot be larger than the width of the space at the small end of the teeth, it is necessary to set the cutter off-center and to rotate the blank to make spaces for the correct width at the large end of the teeth. The distance to offset the cutter from the central position can be determined accurately by using the following formula:

$$\text{Offset} = \frac{T}{2} - \frac{\text{factor from table}}{P}$$

in which P is the diametral pitch of gear to be cut, and T is the thickness of cutter used, measured at the pitch line.

As an example, say you want to cut a bevel gear with 24 teeth, diametral pitch of 6, 30° pitch cone angle, and 1¼-inch face.

Obtain the factor from Table 4-3. The ratio of the pitch cone radius to the width of the face must be determined. The pitch cone radius is equal to the pitch diameter divided by twice the sine of the pitch cone angle, which is $4 \div (2 \times 0.5) = 4$ inches. As the given face width is $1\frac{1}{4}$ inches, the ratio is $4 \div 1.25$, or approximately $3\frac{1}{4}$ to 1. Thus, for a No. 4 cutter, the factor in the table for a $3\frac{1}{4}$ to 1 ratio is 0.280. The thickness of the cutter at the pitch line is measured by using a vernier gear tooth caliper. The cutter thickness at this depth will vary with different cutters, and even with the same cutter as it is ground away in sharpening. Assuming that the thickness is 0.1745 inch, the offset can be determined by substituting in the formula:

$$\text{Offset} = \frac{T}{2} - \frac{\text{factor from table}}{P}$$

$$= \frac{0.1745}{2} - \frac{0.280}{6} = 0.0406 \text{ inch}$$

Figure 4-17 shows an illustration of offsetting the cutter to mill bevel gears, and Figure 4-18 shows a method of filing to correct the shape of the tooth at the small end. Figure 4-19 shows a setup for milling a bevel gear on the milling machine.

In cutting bevel gears, the blank is adjusted laterally the amount of the offset, and the tooth spaces are milled around the blank. After having milled one side of each tooth, the blank is set over in the opposite direction the same amount from a position central with the cutter and is rotated to line up the cutter with a tooth space at the small end. In Figure 4-17, the table is moved in the right-hand direction, and the blank is brought to the correct position by rotating it in the direction indicated by the arrow. A trial cut is then taken, which will leave the tooth being milled a little too thick. This trial tooth is made the proper thickness by rotating the blank toward the cutter. To test the amount of offset, measure the tooth thickness (with a vernier gear-tooth caliper) at the large and small ends. If the offset is

Figure 4-17   The offset of the cutter and rotation of the gear blank in cutting bevel gears so that the right-hand side of the cutter will trim the left-hand side of the gear tooth, thereby widening the large end of the tooth space.

Table 4-3 Table for Obtaining Set Over for Cutting Bevels Ratio of Apex Distance to Width of Face = Apex/Face

No. of Cutter	3	3¼	3½	3¾	4	4¼	4½	4¾	5	5½	6	7	8
1	0.254	0.254	0.255	0.256	0.257	0.257	0.257	0.258	0.258	0.259	0.260	0.262	0.264
2	0.266	0.268	0.271	0.272	0.273	0.274	0.274	0.275	0.277	0.279	0.280	0.283	0.284
3	0.266	0.268	0.271	0.273	0.275	0.278	0.280	0.282	0.283	0.286	0.287	0.290	0.292
4	0.275	0.280	0.285	0.287	0.291	0.293	0.296	0.298	0.298	0.302	0.305	0.308	0.311
5	0.280	0.285	0.290	0.293	0.295	0.296	0.298	0.300	0.302	0.307	0.309	0.313	0.315
6	0.311	0.318	0.323	0.328	0.320	0.334	0.337	0.340	0.343	0.348	0.352	0.356	0.362
7	0.289	0.298	0.308	0.316	0.324	0.329	0.334	0.338	0.343	0.350	0.360	0.370	0.376
8	0.275	0.286	0.296	0.309	0.319	0.331	0.338	0.344	0.352	0.361	0.368	0.380	0.386

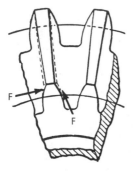

Figure 4-18 The shape of the bevel gear tooth can be corrected at the top on the small end by filing.

correct, the tooth thickness will be the right size at both ends. Then, the cuts can be continued until the gear is finished.

After cutting a bevel gear, the sides of the teeth at the small end should be filed as indicated by the broken lines at *F* in Figure 4-18. A triangular area from the point of the tooth at the large end to the point at the small end and then down to the pitch line and back diagonally to a point at the large end will need filing.

Generating-type gear-cutting equipment is extensively used in the production of bevel gears. A bevel gear tooth that is correctly formed has the same sectional shape throughout its length, but on a uniformly diminishing scale from the large end to the small end. This correct form can be obtained only by using a generating type of bevel gear–cutting machine.

If the bevel gears are too large to be cut by generating equipment (100 inches or more in diameter), they can be produced on a form-copying type of gear planer. A template is used to guide a single cutting tool in the correct path to cut the tooth profile. Since the tooth profile produced by this method is dependent on the contour of the template, it is possible to produce tooth profiles to suit a variety of requirements.

Although generating methods are usually preferred, straight bevel gears are produced by milling in some instances. The milled gears cannot be produced with as much accuracy as generated gears and generally are not suitable for use in high-speed applications or where angular motion must be transmitted accurately. Gears that are to be finished on generating equipment are sometimes first roughed out by milling. Milled gears are used chiefly as replacement gears in certain applications.

Cutting Helical Gears

Helical gears are gears that have the teeth cut along a helical surface. They are usually milled by using standard involute cutters of the arbor-mounted type on the universal knee-and-column milling machine. This permits swiveling the table to the required angle, and the workpiece can be located between centers of a dividing head and tailstock. The operation can be performed on a plain knee-and-column milling machine equipped with a universal

Figure 4-19 Setup for milling a bevel gear. *(Courtesy Cincinnati Milacron Co.)*

milling attachment and a universal dividing head (Figure 4-20). The universal milling attachment permits swiveling the cutter to the required angle of swivel.

Calculations

To determine the angle to which the table must be swiveled for milling a helical gear, the following formula can be used:

$$\text{Tangent of the angle} = \frac{C}{L}$$

in which C is the circumference of work, and L is the lead of the helix. For example, say you want to mill a helix with a lead of 21 inches on a gear blank $1\frac{1}{2}$ inches in diameter.

Figure 4-20   Milling a helical gear on a plain knee-and-column milling machine equipped with a universal milling attachment and a universal dividing head. *(Courtesy Cincinnati Milacron Co.)*

Substituting in the formula,

$$\text{Tangent of the angle} = \frac{4.712}{21} = 0.2244$$

On reference to a table of tangents, an angle of 12°39′ (12.64661929°) is shown to have a tangent of 0.2244. Therefore, the table angle is 12°39′ (12.64661929°).

When a helical gear is rolled on a plane surface, the traces of the teeth line up in equally spaced parallel lines (Figure 4-21). In the illustration, the axial distance Pa between consecutive teeth is the axial pitch, Pn is the normal pitch, and Pc is the circular pitch. The lead of the helix L, the diameter of the cylinder or gear blank D, and the helix angles E and C are also shown.

Helical gears with the shafts at *right angles* to each other have different helix angles; each angle is the complement of the other. When the shafts are *parallel* (Figure 4-22), the helix angle is the same for both gears. If the helical gears have shafts that are at an *angle less than* 90°, the sum of the helix angles is equal to the shaft angle.

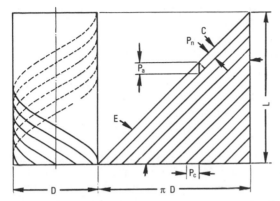

Figure 4-21 Axial, circular, and normal pitch of equally spaced helical gear teeth.

Figure 4-22 Power inspection of helical gears with shafts parallel on a Gleason Universal Tester. *(Courtesy Illinois Gear)*

Setup

Of course, the proper change gears must be selected for connecting the dividing head to the lead screw of the table, so that the workpiece will rotate as the machine table is moved horizontally (see Chapter 3).

The change gears must be placed properly to obtain the correct ratio (see Chapter 3). To generate a left-hand helix, an idler must be introduced in the gear train to reverse the direction of rotation. The driver and driven gears must be placed with care to cut the desired helix.

The number of the cutter to be used in milling a helical gear is also important. The calculated number of teeth to be used in selecting the proper cutter number can be found by dividing the actual number of teeth in the gear by the curve of the cosine of the helix angle.

The gear blank should be mounted on a mandrel, which in turn is mounted between centers on the dividing head. The gear blank is rotated slowly as the table advances at an angle to the axis of the cutter.

The center of the cutter should be directly above the axis of the mandrel carrying the gear blank and the arbor on which the cutter is mounted. The cutter should be centered directly above the axis of the mandrel before swiveling the table to the angle of the helix that is to be machined.

Double helical herringbone gears are often used in parallel shaft transmissions because the opposing helices with the overlapping tooth action provide a smooth, continuous action and freedom from side thrust. These gears are useful in high-speed transmissions (such as marine reduction gears) and in connection with turbine and electric motor drives. Figure 4-23 shows a line of continuous-tooth herringbone gear generators. Some of the gears and gear blanks are shown in the foreground.

Cutting Rack Teeth

Because the pitch line of the rack is a straight line, the base circle (which is parallel) becomes a straight line. The involute of the base circle is a straight line that is perpendicular to the line of action or obliquity. Therefore, the involute rack teeth have straight sides from the bottom of the tooth to the pitch line. Since involute racks are to mesh with pinions that have a relatively small number of teeth, the upper part of each tooth face of the rack is rounded off through a distance equal to one-half the addendum by an arc whose radius is equal to 2.1 inches divided by the diametral pitch. This avoids interference of the teeth. Rack motion is reciprocating, with the rotation of the pinion changing direction periodically. Figure 4-24 shows the involute rack and pinion.

Figure 4-23   A line of continuous-tooth herringbone gear generators.

Cutter Selection

A No. 1 spur gear cutter of the required diametral pitch is commonly used to mill rack teeth. The rack can be compared to a spur gear that has been straightened out and fastened to a flat surface. The center-to-center distance of the rack teeth must equal the circular pitch of a mating gear. The No. 1 spur gear cutter is intended for spur gears varying from 135 teeth to a rack. The depth of tooth is calculated in the same manner as for spur gear teeth.

Setup

Rack teeth (either straight or inclined with respect to the rack blank) can be milled on the milling machine by means of the rack-cutting

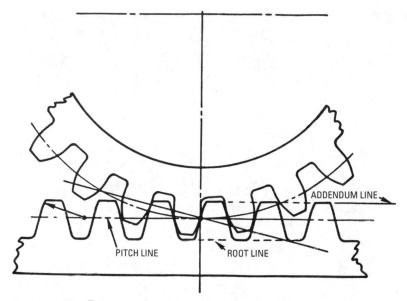

Figure 4-24   Diagram of a rack-and-pinion gear in mesh. *(Courtesy Illinois Gear)*

and rack-indexing attachments. A rack with inclined teeth is known as a spiral rack and can be cut only on universal milling machines (Figure 4-25), since the table can be swiveled to the required angle.

The rack blank is held in a rack vise that is clamped to the table of a universal milling machine (Figure 4-26). Then, the worktable is swiveled in a clockwise direction to a required angle (23°37′ or 23.616666667°). This places the rack blank so that the rack teeth are parallel to the saddle cross-feed and to the milling cutter mounted on the *rack-cutting attachment* (see Figure 4-26).

The machine table moves parallel to the linear pitch of the rack teeth. Thus, to space the teeth at the normal pitch, it is necessary to index each tooth into position by moving the table a distance equal to the linear pitch. The *rack-indexing attachment* (see Figure 4-26) is used to perform the indexing operation. The cutter is mounted on a rack-milling attachment that places the cutter parallel to the cross-feed.

In milling the spiral rack (see Figure 4-25), the workpiece is set vertically by hand-feeding the knee upward until the workpiece contacts the cutter. Then, the dial on the knee adjustment is set at "zero" reading. After clearing the cutter, the knee is raised a distance equal to the whole depth of the tooth to obtain the required depth of cut.

Figure 4-25   Setup for milling the teeth of a spiral rack. *(Courtesy Cincinnati Milacron Co.)*

The workpiece can be located longitudinally by moving the table and saddle until it lightly contacts one side of the cutter. Then, the dial of the table lead screw is set to "zero" reading. After clearing the workpiece from the cutter, the table is moved in a right-hand direction a distance approximately equal to one-half the cutter thickness. This makes the setup ready for milling the rack teeth by feeding the saddle inwardly and returning it to the starting position after each pass.

The next tooth is indexed into position by turning the crank of the rack-indexing attachment until the index plate has made one complete turn. The dimensions and spacing of the first few teeth should be checked. The procedure can then be repeated several times until the required number of rack teeth has been milled.

Cutting Worm and Worm Wheel Teeth

Thread-milling cutters and *hobs* can be used on the milling machine to cut worm wheel teeth. This type of gearing is used in drive arrangements to obtain a reduction in speed ratio between the worm and worm wheel (Figure 4-27). The ratio of the drive is independent of the pitch diameters of the worm and worm wheel. The worm is the driver, and the worm wheel is the driven gear.

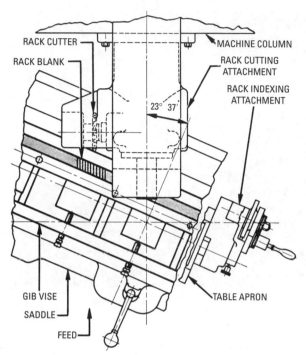

Figure 4-26 Setup for milling rack teeth, using the cross-feed of the machine. The milling cutter is mounted on a rack-cutting attachment that places the cutter parallel to the motion of the saddle.

(Courtesy Cincinnati Milacron Co.)

The *worm* is a screw with either a single thread or multiple threads. The form of the axial cross section is the same as that of a rack. The teeth of the *worm wheel* have a special form that is required to provide proper conditions for meshing with the worm.

The ratio of the drive is equal to the number of teeth in the worm wheel if the worm has a single thread. The drive ratio decreases as the number of threads in the worm is increased, with a constant number of teeth in the worm wheel. For worms with double and quadruple threads, for example, the drive ratio is one-half and one-fourth, respectively, of the number of teeth in the worm wheel.

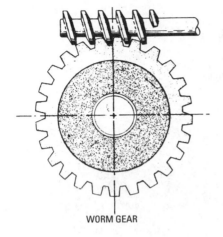

WORM GEAR

Figure 4-27 A typical worm and worm wheel. *(Courtesy Cincinnati Milacron Co.)*

The worm can be milled on a milling machine by means of thread-milling cutters. The worm is held between centers of a dividing head and tailstock. The cutter is mounted on the spindle of a universal milling attachment and swiveled to the helix angle of the worm threads. This setup can be made on a universal knee-and-column milling machine.

Other methods of machining worm threads include the following:

- Milling with a disc-shaped cutter
- Hobbing on a regular gear-hobbing machine
- Generating on a worm thread generator
- Cutting in a lathe if the lead angle is not too great
- Grinding to the finished size, after milling and hardening the worm (the latter is the preferred method if a hardened worm is desired)

Two operations are required to cut the teeth on a worm wheel: *gashing* the teeth and *hobbing* the teeth to the finished size and shape.

In the gashing operation, the gear teeth are *roughed* with an involute gear cutter that has the same pitch and diameter as the worm. After aligning the gear blank on the center of the gashing cutter both crosswise and longitudinally, the table of the machine is swiveled to the gashing angle in a counterclockwise direction for a right-hand worm. The gashing operation is then performed by

Figure 4-28   Setup for gashing the teeth in a worm wheel.
(Courtesy Cincinnati Milacron Co.)

feeding the work vertically to the depth of the teeth, leaving enough stock for the finishing operation (Figure 4-28).

If the gashing angle is not given, it can be calculated from the known value of the lead and pitch diameter of the worm. Tables are usually provided to give the gashing angles for worm wheels for a variety of worms having standard diameters and leads.

The worm wheel is held between centers for the hobbing operation, but it is free from the driving dog of the dividing head. Thus, the hob drives the wheel while the teeth are being cut. The axis of the worm wheel is at right angles to that of the worm. Therefore, it is necessary to set the table of the universal milling machine in the usual straight position so that the axis of the worm wheel is at right angles to the arbor on which the hob is mounted (Figure 4-29).

The table of the machine is locked in position to prevent its moving while the teeth are being hobbed. The workpiece must be adjusted so that the hob centers over the rim of the worm wheel. Then, the work is raised gradually until the correct depth is obtained. If it is necessary to remove a large amount of stock or an exceptional finish is required, the worm wheel can be passed under the hob several times, bringing it to the final depth for the last revolution.

Figure 4-29   Setup for hobbing the teeth of a worm gear.
(Courtesy Cincinnati Milacron Co.)

Summary

A disc, or wheel, that has teeth around its periphery for the purpose of providing a positive drive by meshing the teeth with similar teeth is called a gear. The spur gear is the simplest type and is the most commonly used. The spur gear has straight teeth that are cut parallel with the axis of rotation of the gear body. All other forms of gears (such as bevel, helical, worm, and worm wheels) are modifications of the spur gear.

Various forms are used in connection with gears and gear cutting, such as these terms: pitch circle, pitch point, face, addendum circle, dedendum circle, circular pitch, chordal pitch, pitch diameter, and pressure angle.

The pitch circle is the reference circle of measurement. The pitch point is the surface of teeth between the pitch circle and the top of the tooth. The addendum circle is the circle that passes through the top of the gear teeth and is the same as the outside diameter of the gear.

Thread-milling cutters and hobs can be used on the milling machine to cut worm wheel teeth. This type of gearing is used in drive arrangements to obtain a reduction in speed ratio between the worm and worm wheel.

Helical gears are gears that have the teeth cut along a helical surface and are usually milled by using involute cutters of the arbor-mounted type of the universal knee-and-column milling machine.

The addendum is the height of the tooth above the pitch circle, and the dedendum is the depth of the tooth space below the pitch circle. Circular pitch is the distance from the center of one tooth to the center of the adjacent tooth when measured on an arc of the pitch circle.

Review Questions

1. What is a spur gear?
2. What is the pitch point?
3. The _____ gear teeth differ with the system used to generate them.
4. What is the cycloidal system?
5. Describe the epicycloid curve.
6. What is the hypocycloid curve?
7. The circular pitch system is commonly used if the gear teeth are larger than _____ diametral pitch.
8. Fill in the number of teeth to a rack (standard spur gear cutters):

   No. 1 _____ teeth

   No. 2 _____ teeth

   No. 3 _____ teeth

   No. 4 _____ teeth

   No. 5 _____ teeth

   No. 6 _____ teeth

   No. 7 _____ teeth

   No. 8 _____ teeth

9. To select the proper cutter, it is necessary to know the number of teeth and either the diametral or the _____ pitch.
10. For what is a vernier gear-tooth caliper used?
11. Bevel gears are _____ gears and are used to connect shafts that have intersecting axes.
12. What is the most commonly used type of all the bevel gears?
13. What are Zerol bevel gears?

14. Spiral bevel gears have _____ oblique teeth.

15. What type of gear is useful where hardened gears of high accuracy are required?

16. What type of gear resembles a spiral bevel gear?

17. Define the following terms:

  a. Pitch line

  b. Pitch angle

  c. Pitch cone radius

  d. Addendum

  e. Dedendum

  f. Force angle

  g. Pitch diameter

  h. Cutting angle

18. In cutting _____ gears, the blank is adjusted laterally the amount of the offset, and the tooth spaces are milled around the blank.

19. Generating-type gear-cutting equipment is extensively used in the production of _____ gears.

20. Helical gears are gears that have the teeth cut along a _____ _____ surface.

21. What gears are useful in high-speed transmissions?

22. A No. 1 spur gear cutter of the required _____ pitch is commonly used to mill rack teeth.

23. What is a rack-indexing attachment?

24. Thread-milling cutters and hobs can be used on the milling machine to cut _____ wheel teeth.

25. What are the two operations required to cut the teeth on a worm wheel?

Chapter 5

Cams and Cam Design

A cam is a mechanical device used to convert *rotary* motion into *varying-speed reciprocating* motion. This basic action results when an irregularly shaped disc revolves on a shaft and imparts irregular motion to a follower. Cams, in general, can be divided into two classes: *uniform motion cams* and *uniformly accelerated motion cams*.

Cam Principles

The motion of the cam is transmitted to a follower, which remains in contact with the acting surface as the cam revolves. The follower corresponds to the plunger or lifter of the valve mechanism (Figure 5-1).

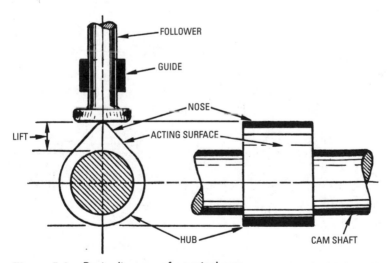

Figure 5-1 Basic diagram of a typical cam.

Cams are made in two basic shapes: plate cams and cylindrical cams. In general, the follower of a plate cam operates perpendicular to the axis of the camshaft while the follower of the cylindrical cam oscillates parallel with the axis of the camshaft (Figure 5-2).

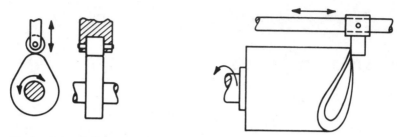

Figure 5-2 Follower motion in respect to cam action.

Cam followers may have a variety of shapes. Figure 5-3 shows the following four common shapes: knife-edge (or pointed), roller, flat-faced, and spherical-faced. Follower motion is assured by the use of gravity, spring action, or operating in a groove. A groove cam has the advantage of being able to apply force in opposite directions (Figure 5-4).

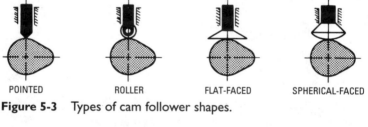

POINTED   ROLLER   FLAT-FACED   SPHERICAL-FACED

Figure 5-3 Types of cam follower shapes.

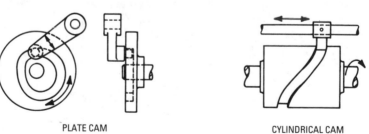

PLATE CAM   CYLINDRICAL CAM

Figure 5-4 Groove cam applies force in opposite directions.

Uniform Motion Cams

If a body moves through equal spaces in equal intervals of time, it is said to have *uniform motion* (that is, its velocity is constant). The uniform motion cam moves the follower at the same rate of speed throughout the stroke from the beginning to the end. This means that, if the movement is rapid, there is a distinct shock at the beginning and end of the stroke because the movement is started from "zero" to the top speed of the uniform motion and is stopped just as abruptly. Therefore, it is important to construct cams in such a way that these sudden shocks are avoided either in starting the motion or in reversing the direction of motion of the follower. These cams are suitable for machinery operating at a slow rate of speed.

Uniformly Accelerated Motion Cams

The cam best suited for high speeds is one in which the speed is slow at the beginning of the stroke, accelerated at a uniform rate until maximum speed is attained, and then uniformly retarded until the follower is stopped (or nearly stopped) when the reversal takes place. This type of cam is called a *uniformly accelerated* (or *retarded*) *motion cam*.

Harmonic Motion

Figure 5-5 shows this form of varying or uniformly accelerated motion. If a point A, as indicated by the arrow, travels around the circumference of a circle at uniform velocity, and another point B travels across the diameter of the circle at a variable velocity so that it is always at a point where a perpendicular from point A would intersect the diameter, then the point B increases from "zero" at the starting point C until it reaches the center point O. From that point, the velocity decreases to "zero" as it reaches the point D at the end of its travel. The harmonic motion cam is satisfactory for machinery operating at moderate speeds.

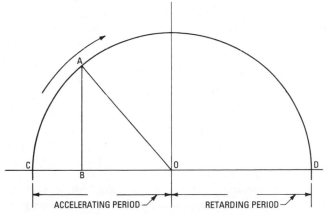

Figure 5-5 Diagram showing harmonic motion.

Gravity Motion

In this uniformly accelerated (and retarded) motion, the rate of acceleration or retardation bears the same ratio to the speed as the acceleration or retardation produced by gravity—hence its name. A body falling from rest travels about 16 feet during the first second. During the next second, its velocity increases by 32 feet/second, making 48 feet the distance covered during that second. During each succeeding second, the velocity increases by 32 feet/second. The increase in velocity is in the ratio of 1, 3, 5, 7, 9, and so on. This ratio can be used in laying out a gravity motion cam. The gravity motion cam is best for machinery operating at high speeds.

How a Cam Operates

Figure 5-6 shows tangential cam action in opening and closing a valve. The cam and valve positions are shown for each 22.5° during 180°, or one-half of a revolution. The successive cam positions are indicated by the numerals 1, 2, 3, and so on. The corresponding valve positions are also indicated by the same numbers. In cam positions 1 and 2, the follower is in contact with the cam hub. The valve remains seated in these positions. As can be seen in Figure 5-6, the valve remains seated until the straight portion of the cam's acting surface attains a horizontal position, as indicated by the dotted outline A.

When the cam reaches position 3, it has pushed the follower and valve upward a distance a; the valve position is shown directly above. Similarly, for cam positions 4, 5, 6, and so on, the valve is upward at the distances b, c, d, and so on, respectively, from its seat. The valve reaches its closing point between positions 7 and 8.

The curve at the top of the diagram indicates the movement of the valve for 180°, or one-half of a revolution. The curve indicates that the cam is a type that has a varying motion (that is, the motion is accelerated in its opening and retarded in its closing movements).

The points of opening and closing the valves are the points where the acting surface is tangent to the hub surface (Figure 5-7). In the tangential cam, the acting surface between the nose and hub is straight. Perpendiculars to the straight surface indicate the valve positions as determined by the positions of the ends of the valve stems. In the rotation of the cam, these two points cause the valve to open and close as they come in contact with the follower. The arc ARB is the rest period in which the valve remains closed.

When the follower motion is parallel to the camshaft, it is necessary to use some form of a cylindrical cam. This usually consists of a roller follower operating in a groove cut in the rotating cylinder

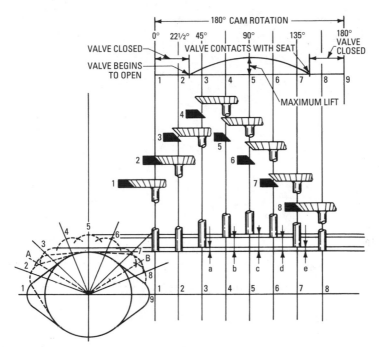

Figure 5-6   Cam action needed when opening and closing a valve.

cam. The groove is cut according to the amount of displacement and the type of motion desired (Figure 5-8).

Cam Design
Cam action has its limits. Under certain conditions, the result can be a noisy and fast-wearing mechanism. Some basic or fundamental facts should be considered in cam design.

Displacement Diagrams
Since the motion of the follower is of primary importance, its rate of speed and its various positions should be carefully planned in a displacement diagram. A displacement diagram is a curve showing the displacement of the follower as ordinates erected on a baseline that represents one revolution of the cam (Figure 5-9). These three diagrams illustrate the various cam motions while Figure 5-10 shows a displacement diagram of a cam combining all three motions. Notice that the uniform motion has been modified to reduce shock.

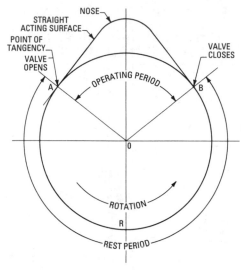

Figure 5-7 The determination of points of opening and closing in cam rotation.

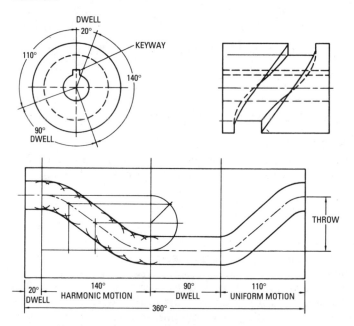

Figure 5-8 Layout of cylindrical cam.

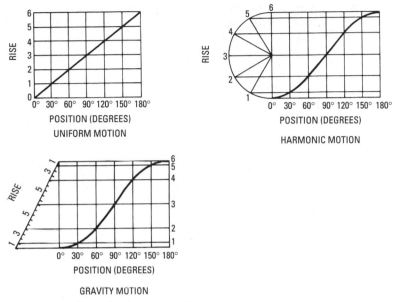

Figure 5-9 Cam displacement diagrams.

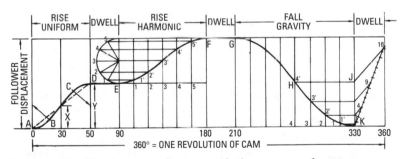

Figure 5-10 Displacement diagram with three types of motion.

Regardless of the desirability of quick action in valve opening and closing, the opening and closing faces should meet the hub circle tangentially (that is, there should be no abrupt change in direction of the acting surface). This is especially important on the opening face because the tension of the spring opposes a sudden starting action of the valve when it begins to open. Also, with the steeper opening face, the lateral thrust on the follower is greater, which increases friction and wear.

In a uniform motion cam, the contour of the action surface moves the follower at the same rate of speed from beginning to end of the valve movement. Figure 5-11 shows the layout of a uniform motion cam.

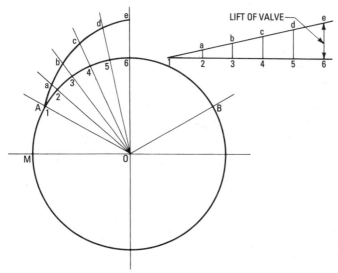

Figure 5-11 The layout of a cam having a uniform motion.

Valve lift is indicated in the upper part of the illustration (see Figure 5-11). If a line is divided into a number of equal parts and if perpendiculars are drawn from the points 1, 2, and so on to intersect the inclined line *le*, movement of the valve is indicated by the distances 2*a*, 3*b*, and so on for equal arcs of cam rotation (1-2, 2-3, and so on).

In the lower part of Figure 5-11, the horizontal and vertical axes are laid out, and the lines *OA* and *OB* are constructed so that the obtuse angle *AOB* represents the operating angle of the cam. Then, a circle having a radius equal to that of the cam hub is described, and the arc 1-6 is divided into the same number of equal parts as the line 1-6 was divided into in the upper part of the illustration. The radii of the circle are extended indefinitely through the points 1, 2, and so on, far enough to lay off the distances 2*a*, 3*b*, and so on, equal to the corresponding distances in the upper part of the illustration.

Then, the curve *abcde* can be drawn through the points on the radii to represent one-half of the acting surface of the cam. The other one-half of the curve is identical and can be obtained by using the same construction method.

The chief disadvantage of the uniform motion cam is that valve motion begins at full speed and stops in the same abrupt manner, which is a distinct shock to the valve gear in starting and stopping actions. This is illustrated in Figure 5-11. As point 1 of the cam contacts the follower, the follower immediately moves at full speed (that is, without a gradual acceleration period). It is common practice to modify this type of action to reduce shock, as shown in Figure 5-10, where the dashed line represents the true uniform motion and the solid line represents the modified uniform motion. This type of cam shape is unsatisfactory for automobile engines because of the high speed at which they are run.

Design for Gas Engines

Layout of a cam for operating a valve under specified conditions is not complicated. The rapidity with which the valves open and close is determined by the slope of the valve opening and closing faces on the cam. Steep opening and closing faces on the cam provide quick movement of the valves. The following example can be used to illustrate the design of a uniformly accelerated motion cam.

The cam is to be designed to operate an intake valve. The following conditions are specified: lift, $^{11}/_{32}$ (0.34375) inch; cam hub circle, $1\frac{1}{2}$ (1.50) inches; nose arc, $^{5}/_{16}$ (0.3125) inch; admission period, 240°. The cam is to provide quick acceleration in the opening and closing actions of the valve.

As an admission period occupies 240° of crankshaft rotation, it will occupy only 120° of camshaft rotation because the camshaft speed is one-half that of the crankshaft (240° ÷ 2 = 120°). Figure 5-12 shows the design of the cam with curved opening and closing faces for operation of the intake valve.

First, draw the horizontal and vertical axes (see Figure 5-12). Draw lines *Aa* and *Bb* through the center point O so that the angle *AOB* is equal to 120° and is bisected by the vertical axis ($\angle AOM$ = 60° and $\angle BOM$ = 60°). Describe the cam hub circle [1½ (1.50) inches], and lay off the lift [$^{11}/_{32}$ (0.34375) inch] from point m. With the center of the vertical axis *OM*, describe the nose arc [with radius of $^{5}/_{16}$ (0.3125) inch] through point n. On line *Aa*, by trial, find the center of a circle that is tangent to both the cam hub circle and the nose arc. Describe the arc connecting them to give the contour of the opening face. The contour of the closing face is identical and can be obtained by the same method of construction. This is frequently referred to as a *circular arc cam* since the surface consists entirely of tangent arcs of circles. It has the advantages of being comparatively easy to manufacture.

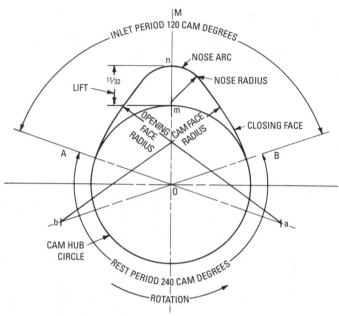

Figure 5-12   The design of a cam having uniformly accelerated motion.

Another example can illustrate the design of a tangential cam for intermittent motion. Using the same lift, admission period, and cam hub diameters as in the preceding example, the problem differs in this example (Figure 5-13) as follows:

- The nose arc is described from the center point O of the cam; and
- The opening and closing faces are tangent lines that are tangent to the hub circle at the points of intersection of the lines *Aa* and *Bb* with the hub circle.

It should be noted that the action on the follower is very abrupt at the points f and f' (see Figure 5-13). This can be avoided by connecting the tangents and nose arc with relief arcs, as shown. The amount of relief (that is, the acceleration and retardation periods) depends on the radius of the relief arcs. Solid lines are used to show the arcs with smaller relief, and dotted lines are used to show the arcs with larger relief. This cam is referred to as a *tangential cam* since the opening and closing faces are tangent to the hub circle and the relief arcs.

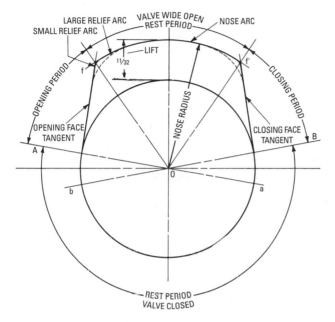

Figure 5-13   The design of a tangential cam for intermittent motion.

An *over-travel cam* is sometimes used to give the valve more opening than is necessary for proper distribution of the charge. Over-travel is used to increase acceleration and retardation and obtain relief at reversal of movement. Valve lift can vary with the manufacturer.

The design of cams for exhaust valves is different from that of intake valves. The cams for exhaust valves have a more flattened nose so that a period of maximum opening is provided for the free escape of the burned gases.

Figure 5-14 shows a comparison of sharp-nosed cams and broad-nosed cams. When the cams are in the positions M and S, they are at the points where the valves begin to open. At the positions L and F, they are fully open. As noted in the illustration, the broad-nosed cam provides greater acceleration in opening the valve.

Design for Automatic Screw Machines

The Brown & Sharpe No. 2 high-speed machine is used here to illustrate typical cam design. (The reader may wish to refer to Chapter 15 of *Audel Machine Shop Basics: All New Fifth Edition*, which describes the setup of this machine.) A typical workpiece,

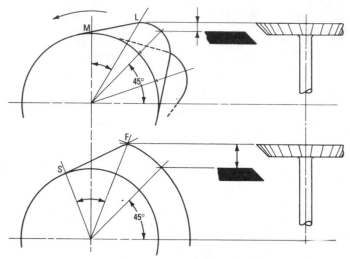

Figure 5-14 A comparison of acceleration on a sharp-nosed cam (top) and a broad-nosed cam (bottom). The broad-nosed cam provides greater acceleration in the opening of the valve.

shown in Figure 5-15, is also used. This particular workpiece has a turned diameter, a formed head, and a rounded end. The necessary steps in cam design are as follows:

1. Decide on the method to be used in doing the job, tools required, and order of operations.

2. Determine spindle speed.

3. Calculate throws of the cam lobes and the spindle revolutions required for the cutting operations.

4. Overlap the operations wherever possible.

5. Figure spindle revolutions required for idle movements.

6. Provide clearance space if necessary.

7. Find total estimated spindle revolutions required to finish a piece. Then, select the actual revolutions available with regular change gears that come nearest the estimated number.

8. Readjust the estimated spindle revolutions to total the actual number available on the machine.

9. Calculate the hundredths of cam surface required for each operation and idle movement.

As an illustration, we can assume that the turned diameter on the workpiece (see Figure 5-15) requires a good finish, the following

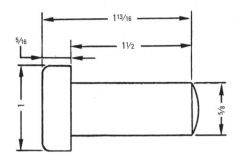

Figure 5-15 Workpiece used to illustrate typical cam design for an automatic screw machine.

operations being necessary:

1. Rough turn to within 0.010 inch of the finished diameter.
2. Turn with a balance-turning tool to within 0.010 inch of the finished length.
3. Finish turn with a box tool to the finished diameter.
4. Advance the cross slides to form the head and sever the workpiece from the bar.
5. Finish under the head with the forming tool, removing the 0.010 inch of length left by the turning tools.
6. Round the end with the cutoff tool for the next workpiece.

A worksheet (Figure 5-16) can be used to assemble data for making the cam. This is not justified unless there is a large volume of work to be performed. The cam design worksheet gives all the data needed to do the job with all the operations listed in the order in which they are performed.

Lever templates (Figure 5-17) can be used to calculate close timing in cam design. It is necessary to know the relative positions of one or both of the cross slides when the turret is advancing, retreating, or at its extreme inward position. Likewise, it is essential at times to know the position of the turret slide relative to one or both cross slides. The templates are useful in determining the exact relative positions of the slides on the drawing without having to resort to figures or calculations. They are useful in determining the amount of space to allow for tool clearance or determining points that correspond on the lead cam and cross-slide cams so that the movements of the turret slide and cross slides can be synchronized when their combined action is required to feed a tool (as in swing tool work).

The templates pivot at the center of the cam drawing or trial sketch on a pin inserted through the hole in a button at the end of the long arm. They are made of clear material so that the lines of

CAM DESIGN WORK SHEET

PART NAME OR NO.... Example 1 .... MATERIAL BRASS ....

SURFACE FEET FOR STOCK ....

" " " THREAD ....

" " " DRILL ....

" " " TURN ....

SPINDLE (FORWARD.... 3000 ....

R/MIN (BACKWARD ....

SECONDS....12 ....

GROSS PRODUCTION PER HOUR..300 ....

MADE ON.. NO. 2 High Speed.................

Order of Operations	Throw	Feed	For Each Operation	After Deducting for Operations Overlapped	Readjusted to Equal Revolutions obtainable with Regular Change Gears	For Spindle Revolutions in Preceding Column	For Operations That Are Overlapped
				Spindle Revolutions		Hundredths	
Feed Stock to Stop (4" long) ..					30	30	5
Double Index Turret					30	30	5
Rough Turn - No. 22							
Balance Turning Tool	1.500	0.010	150		150	150	25
Double Index Turret					30	30	5
Finish Turn - No. 22B							
Box Tool	1.500	0.010	150		150	150	25
Clear						24	4	3
From - Front Slide .... {	182	0.003	60				10
	010	0.001	18				3
Double Index Turret							
Cut Off - Back Slide	574	0.0035	164		164	186	31
96 Hundredths equal					554		
Estimated Total Revs., if spindle runs 3000 r/min continually					576		
Nearest Actual Spindle Revs. available on Machine						600	100

NOTE—A dimensioned pencil sketch of the piece is often drawn in blank space at top of this sheet.

Figure 5-16 Cam design worksheet.

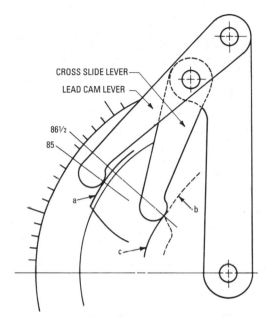

Figure 5-17   Using lever templates in cam design.

the cam drawing can be seen. The centers from which the lead and cross slide levers swing represent the fulcrum centers of the actual levers on the machine. The two levers are the same length as the corresponding levers on the machine, and the rounded portion at their ends is the same diameter as the rolls that run on the cams. If, for example, the lead cam lever is at the 85th division line on lobe *a*, and you want to have the front cross slide begin to guide or feed the tool at that instant, the corresponding hundredth division line on the cross-slide cam can be found by placing the lead cam lever with the roll barely touching the line of lobe *a* at the 85th division line. Then, swing the cross-slide lever downward until it touches the arc *c*, which represents the bottom of the throw for lobe *b*. A radial line from the center point of the cam drawing to the cam circle and passing through the center of the roll on the cross-slide lever shows graphically the hundredth division line on which to start the lobe *b*—at 86.5 hundredths in this instance.

Estimating Spindle Revolutions
In laying out cams, the approximate feeds and speeds can be obtained from Table 5-1. From the table, it can be noted that brass

can be worked at the fastest speed of the machine, which is 3000 revolutions per minute (rpm), as indicated in the table for laying out cams (Table 5-2). It should be noted in Table 5-2 that when threading to a shoulder, the required relative position of the clutch teeth at the time of reversing is obtained by using the table values. This exists when a whole number is obtained by multiplying "time in seconds" by 48. Time in seconds is obtained as follows:

$$\text{Time in seconds} = \frac{\text{Ist gear on } B}{\text{gear on } A} \times \frac{\text{gear on } C}{\text{2nd gear on } B}$$

Care should be taken to select combinations that will mesh but will not interfere.

Table 5-1 Approximate Cutting Speeds and Feeds for Standard Tools

					Material					
					Mild or Soft Steel, 0.10–0.20% Carbon			Tool Steel, 0.80–1.00% Carbon		
	Cut		Brass, Free Cutting		Speed in Surface Feet			Speed in Surface Feet		
Tool	Width or Depth	Diameter of Hole	Feed	Speed in Surface Feet	Feed	Carbon Tools	H.S.S.* Tools	Feed	Carbon Tools	H.S.S.* Tools
Boring tools	0.005				0.008	50	110	0.004	30	60
Box tools— roller rest	$\frac{1}{32}$		0.012		0.010	70	150	0.005	40	75
Chip finishing	$\frac{1}{16}$		0.010		0.008	70	150	0.004	40	75
	$\frac{1}{8}$		0.008	Use maximum spindle speed	0.007	70	150	0.003	40	75
	$\frac{3}{16}$		0.008		0.006	70	150	0.002	40	75
	$\frac{1}{4}$		0.006		0.005	70	150	0.0015	40	75
Finishing	0.005		0.010		0.010	70	150	0.006	40	75
Center drills		Under $\frac{1}{8}$	0.003		0.0015	50	110	0.001	30	75
		Over $\frac{1}{8}$	0.006		0.0035	50	110	0.002	30	75
Cut-off tools, angular			0.0015		0.0006	80	150	0.0004	50	85
circular	$\frac{3}{64}$–$\frac{1}{8}$		0.0035		0.0015	80	150	0.001	50	85
Straight	$\frac{1}{16}$–$\frac{1}{8}$		0.0035		0.0015	80	150	0.001	50	85
Diameter stock under $\frac{1}{8}$ in.			0.002		0.0008	80	150	0.0005	50	85

Table 5-1 (continued)

Tool	Width or Depth	Diameter of Hole	Feed	Speed in Surface Feet	Feed	Carbon Tools	H.S.S.* Tools	Feed	Carbon Tools	H.S.S.* Tools
			Brass, Free Cutting		Mild or Soft Steel, 0.10–0.20% Carbon — Speed in Surface Feet			Tool Steel, 0.80–1.00% Carbon — Speed in Surface Feet		
Button dies							30		14	
Chaser dies						30	40		16	20
Drills	0.02		0.0014		0.001	40	60	0.0006	30	45
Twist cut	0.04		0.002		0.0014	40	60	0.0008	30	45
	1⁄16		0.004		0.002	40	60	0.0012	30	45
	3⁄32		0.006		0.0025	40	60	0.0016	30	45
	1⁄8		0.009		0.0035	40	75	0.002	30	60
	3⁄16		0.012		0.004	40	75	0.003	30	60
	1⁄4		0.014		0.005	40	75	0.003	30	60
	5⁄16		0.016		0.005	40	75	0.0035	30	60
	3⁄8		0.016		0.006	40	85	0.004	30	60
	1⁄2		0.016		0.006	40	85	0.004	30	60
	5⁄8		0.016		0.006	40	85	0.004	30	60
					0.0009			0.0006		
Form tools— circular	1⁄8–1⁄4		0.002		0.0008			0.0005		
	3⁄8–1⁄2		0.0015 0.0012	Available on machine	0.0007 0.0006	80	150	0.0004	50	85
	5⁄8–3⁄4		0.001		0.0005	80	150	0.0003	50	85
	1		0.001		0.0004	80	150			
Hollow mills (turned diameter under 5⁄32 in.)	1⁄32		0.012		0.010	70	150	0.008	40	85
	1⁄16		0.010		0.009	70	150	0.006	40	85
Balance turning tools (turned diam. over 5⁄32 in.)	1⁄32		0.017		0.014	70	150	0.010	40	85
	1⁄16		0.015		0.012	70	150	0.008	40	85
	1⁄8		0.012		0.010	70	150	0.008	40	85
	3⁄16		0.010		0.008	70	150	0.006	40	85
			0.009		0.007	70	150	0.0045	40	85

(continued)

114 Chapter 5

Table 5-1 (continued)

Tool	Cut Width or Depth	Cut Diameter of Hole	Brass, Free Cutting Feed	Brass, Free Cutting Speed in Surface Feet	Mild or Soft Steel, 0.10–0.20% Carbon Feed	Carbon Tools	H.S.S.* Tools	Tool Steel, 0.80–1.00% Carbon Feed	Carbon Tools	H.S.S.* Tools
Knee tools	1/64–1/32				0.010	70	150	0.010	40	85
Knurl tools					0.012			0.008		
Turret	On		0.020		0.015	150		0.010	105	
	Off		0.040		0.030	150		0.025	105	
Side or swing			0.004		0.002	150		0.002	105	
			0.006		0.004	150		0.003	105	
			0.005		0.003	150		0.002	105	
Top			0.008		0.006	150		0.004	105	
			0.001		0.0008	70	150	0.0005	40	80
Pointing & facing tools			0.0025		0.002	70		0.0008	40	80
Reamers & bits	.003 to .004	1/8 or less	0.010		0.008	70	150	0.006	40	60
			0.007	Available on machine	0.006	70	105	0.004	40	60
	.004 to .008	1/8 or over	0.010		0.010	70	105	0.006	40	60
								0.008	40	60
Recessing tools, end cut			0.001		0.0006	70	150	0.0004	40	75
			0.005		0.003	70	150	0.002	40	75
Inside cut	1/16–1/8		0.0025		0.002	70	105	0.0015	40	60
			0.0008		0.0006	70	105	0.0004	40	60
Swing tools, forming	1/8–1/4		0.002		0.0007	70	150	0.0005	40	85
			0.0012		0.0005	70	150	0.0003	40	85
	3/8–1/2		0.001		0.0004	70	150	0.0002	40	85
			0.0008		0.0003					
Turning straight	1/32		0.008		0.006	70	150	0.0035	40	85
	1/16		0.006		0.004	70	150	0.003	40	85
	1/8		0.005		0.003	70	150	0.002	40	85
	3/16		0.004		0.0025	70	150	0.0015	40	85
Taps						25	30		12	15
Swing tools										

*H.S.S. is High-Speed Screw Machine.
Taper turning is the same as straight turning, but the feed is taken slowly enough for the greatest depth of cut.

Table 5-2 No. 2 Automatic High-Speed Screw Machine (With 36 Spindle Speeds)

Feed Gears

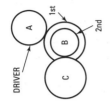

Spindle Speeds

3000	2460	2110	1790	1530	1250	1090	895	765	650	560	455
1070	880	750	640	545	445	390	320	275	230	200	160
645	525	450	385	330	270	235	190	165	140	120	100

Table for Laying Out Cams—Driving Shaft 240 rpm

Revolutions at Maximum Speed to Feed Stock or Index Turret—½ Second

25	21	18	15	13	10	9	8	7	6	5	4	Time to Make One Piece (sec.)	Gross Product per Hour*	Gear on A	1st Gear on B	2nd Gear on B	Gear on C	Hundredths of Cam Surface to Feed Stock

Revolutions of Spindle at Maximum Speed to Make One Piece

25	21	18	15	13	10	9	8	7	6	5	4	Time to Make One Piece (sec.)	Gross Product per Hour*	Gear on A	1st Gear on B	2nd Gear on B	Gear on C	Hundredths of Cam Surface to Feed Stock
150	123	105	90	77	63	54	45	38	33	28	23	3	1200	80	32	80	40	17
175	144	123	104	89	73	64	52	45	38	33	27	3½	1028	80	32	72	42	15
188	154	132	112	96	78	68	55	48	41	35	28	3¾	960	80	36	72	40	14
200	164	141	119	102	83	73	60	51	43	37	30	4	900	80	32	72	48	13
208	171	146	124	106	87	76	62	53	45	39	32	4⅛	864	72	40	80	40	12
225	185	158	134	115	94	82	67	57	49	42	34	4½	800	80	32	72	54	12
233	191	164	139	119	97	85	70	60	51	44	35	4⅔	771	72	48	90	42	11
250	205	176	149	128	104	91	74	64	54	47	38	5	720	80	32	72	60	10

Table 5-2 (continued)

Table for Laying Out Cams—Driving Shaft 240 rpm

Spindle Speeds												Time to Make One Piece (sec.)	Gross Product per Hour*	Gear on A	1st Gear on B	2nd Gear on B	Gear on C	Hundredths of Cam Surface to Feed Stock
3000	2460	2110	1790	1530	1250	1090	895	765	650	560	455							
1070	880	750	640	545	445	390	320	275	230	200	160							
645	525	450	385	330	270	235	190	165	140	120	100							
Revolutions at Maximum Speed to Feed Stock or Index Turret—½ Second																		
25	21	18	15	13	10	9	8	7	6	5	4							
Revolutions of Spindle at Maximum Speed to Make One Piece																		
275	226	193	164	140	115	100	82	70	60	51	42	5½	655	80	32	84	77	10
300	246	211	179	153	125	109	90	77	65	56	45	6	600	80	32	60	60	9
325	267	229	194	166	135	118	97	83	70	61	49	6½	554	80	32	72	78	8
350	287	246	209	179	146	127	104	89	76	65	53	7	514	80	32	60	70	8
400	328	281	239	204	167	145	119	102	87	75	61	8	450	80	32	60	80	7
450	369	316	269	230	188	163	134	115	98	84	68	9	400	80	32	48	72	6
500	410	352	298	255	208	182	149	128	108	93	76	10	360	80	32	48	80	5
550	451	387	328	281	229	200	164	140	119	103	83	11	327	80	32	42	77	5
600	492	422	358	306	250	218	179	153	130	112	91	12	300	80	32	40	80	5
650	533	457	388	332	271	236	194	166	141	121	99	13	277	80	32	36	78	4
700	574	492	418	357	292	254	209	179	152	131	106	14	257	80	32	36	84	4
750	615	527	448	383	313	272	224	191	163	140	114	15	240	60	60	80	80	4
875	718	615	522	446	365	318	261	223	190	163	133	17½	205	60	60	72	84	3
1000	820	703	597	510	417	363	298	255	217	187	152	20	180	60	60	54	72	3
1125	923	791	671	574	469	409	336	287	244	210	171	22½	160	60	60	48	72	3
1250	1025	879	746	638	521	454	373	319	271	234	190	25	144	60	60	48	80	3

1375	1128	967	821	701	573	499	410	351	298	257	208	27½	131	60	60	42	77	3
1500	1230	1055	895	765	625	545	448	383	325	280	227	30	120	40	80	60	60	3
1750	1435	1231	1044	893	729	636	522	447	379	327	265	35	103	40	80	60	70	3
2000	1640	1406	1194	1020	834	726	597	510	434	374	303	40	90	40	80	54	72	3
2250	1845	1582	1343	1148	938	817	671	574	488	420	341	45	80	40	80	48	72	3
2500	2050	1758	1492	1275	1042	908	746	638	542	467	379	50	72	40	80	48	80	3
2750	2255	1934	1641	1403	1146	999	821	702	596	514	417	55	65	40	80	42	77	3
3000	2460	2110	1790	1530	1250	1090	895	766	650	560	455	60	60	40	80	40	80	3
3375	2768	2373	2014	1721	1407	1226	1007	861	732	630	512	67½	53	36	72	40	90	3
3750	3075	2637	2238	1913	1563	1362	1119	957	813	701	569	75	48	36	80	40	90	3
4125	3383	2901	2462	2104	1719	1498	1231	1053	894	771	625	82½	44	36	77	35	90	3
4500	3690	3164	2686	2295	1876	1634	1343	1148	976	841	682	90	40	36	84	35	90	3
4875	3998	3428	2909	2486	2032	1771	1455	1244	1057	911	739	97½	37	32	78	36	96	3
5250	4305	3692	3133	2678	2188	1907	1567	1340	1138	981	796	105	34	24	80	40	84	3
5625	4613	3956	3357	2869	2345	2043	1679	1436	1220	1051	853	112½	32	32	90	36	96	3
6000	4920	4219	3581	3060	2501	2179	1790	1531	1301	1121	910	120	30	24	80	40	96	3
6750	5535	4747	4028	3443	2813	2452	2014	1723	1463	1261	1023	135	27	32	72	24	96	3
7500	6150	5274	4476	3825	3126	2724	2238	1914	1626	1401	1137	150	24	24	80	32	96	3
8250	6765	5801	4924	4208	3439	2996	2462	2105	1789	1541	1251	165	22	24	88	32	96	3
9000	7380	6329	5371	4590	3751	3269	2686	2297	1951	1681	1364	180	20	22	88	32	96	3
9750	7995	6856	5819	4973	4064	3541	2909	2488	2114	1821	1478	195	18	24	78	24	96	3
10,500	8610	7384	6266	5355	4376	3814	3133	2680	2276	1961	1592	210	17	24	84	24	96	3
11,250	9225	7911	6714	5738	4689	4086	3357	2871	2439	2102	1706	225	16	24	90	24	96	3
12,000	9840	8438	7162	6120	5002	4358	3581	3062	2602	2242	1819	240	15	24	88	22	96	3

*Net will vary with factory conditions and the character of the work.

Figure 5-18 shows the positions of the tools in the turret and on the cross slide for machining the workpiece shown in Figure 5-15. The turret stop is used on this job.

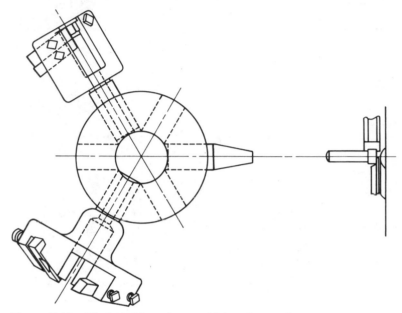

Figure 5-18   Tool positions for machining the workpiece.

To estimate the spindle revolutions for the *balance-turning tools*, several factors must be considered. Length turned is 1.500 inches, less 0.010 inch that is left under the head for the forming tool. Usually, from 0.010 inch to 0.015 inch is added to the length for all turret tools with the exception of taps and dies. This extra length is required to permit the tools to approach the work at required feed and not "jab in" at the beginning of the cut. Thus, (1.500 inches − 0.010 inch) + 0.010 inch = 1.500 inches is the length to be turned or the throw of the cam lobe. The feed for the balance-turning tool depends on the depth of cut (see Table 5-1), which in this instance is:

$$\text{Depth of cut} = \frac{\text{diameter of stock} - \text{finished diameter}}{2}$$

$$= \frac{1.000 - 0.625}{2} = \frac{0.375}{2} = 0.1875$$

Therefore, the feed selected from the table (see Table 5-1) is 0.010 inch per spindle revolution, which is the feed for ³⁄₁₆-inch depth of cut. Dividing 0.010 inch into 1.5 inches (length of cut) gives 150 revolutions (1.5 ÷ 0.010 = 150) of the spindle required for the operation.

For the *box tool*, the turned length is the same as for the balance-turning tool. A feed of 0.010 inch per revolution (see Table 5-1) is used for box tools when finish turning, removing a depth of 0.005 inch, which was the amount left by the balance-turning tool. Therefore, this operation requires 150 spindle revolutions (1.5 ÷ 0.010 = 150).

The *form tool* must advance on the workpiece a distance equal to one-half the difference in the diameter of the stock and the finished diameter or:

$$\text{Advance} = \frac{1.000 \text{ inch} - 0.625 \text{ inch}}{2}$$

$$= \frac{0.375 \text{ inch}}{2} = 0.187 \text{ inch}$$

If a distance of approximately 0.005 inch is allowed for approaching the work with the form tools, then:

0.187 inch + 0.005 inch = 0.192 inch throw

The cutoff tool will have considerable work to do after the forming tool has completed its portion of the work (see Figure 5-15). Therefore, the cutoff tool should be started first. Thus, the forming tool shaves material off the sides of the head only until it approaches full depth. Therefore, it is possible to feed inward until within about 0.010 inch of full depth.

From Table 5-1, a feed of 0.003 inch per revolution is selected for the first 0.182 inch, and 0.001 inch per revolution is selected for the final 0.010 inch of throw:

0.182 inch ÷ 0.003 inch = 61 spindle revolutions

0.010 inch ÷ 0.001 inch = 10 spindle revolutions

If eight spindle revolutions are added to include "dwell" for the last 0.010 inch of travel, a total of 78 spindle revolutions are required for the form tool.

The *cutoff tool* must advance on the workpiece a distance equal to one-half the diameter of stock plus a distance of 0.004 inch to 0.008 inch for approach to the work and a distance large enough to permit the heel of the tool to pass the center and trim the end of the bar, leaving no material. Table 5-3 shows the angles and thicknesses for circular cutoff tools. For example, a cutoff tool for 1-inch brass stock has an angle 0.059 inch in depth. Usually, an allowance of 0.003 inch to 0.005 inch is made for the heel to pass the center point when cutting off a solid bar. The throw of the cutoff tool in this instance is as follows:

   One-half diameter of stock = 0.500 inch (1.000 ÷ 2)
   Approach to work = 0.010 inch
   Depth of angle on tool = 0.059 inch
   Heel of tool past center = 0.005 inch
   Total throw for cutoff tool = 0.574 inch

Table 5-3   Angles and Thicknesses for Circular Cutoff Tools

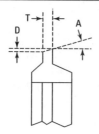

A is 23° when cutting brass, aluminum, copper, silver, and zinc.

A is 15° when cutting steel, iron, bronze, and nickel.

Least thickness used when cutting off into tapped holes is the lead of two and one-half threads plus 0.010″.

Least thickness used when cutting off into reamed holes smaller than ⅛″ diameter is 0.040″.

Thickness used when cutting off tubing is two-thirds T as given below for corresponding diameters of stock.

Thickness used when angles or radii start from outside diameter of tool is governed by varying conditions and determined accordingly.

Diameter of Stock		Thickness (T)	Depth of Angle (D)	
Fractional	Decimal		Brass	Steel
¹⁄₁₆	0.06250	0.020	0.0085	0.0055
³⁄₃₂	0.09375	0.030	0.013	0.008
⅛	0.12500	0.040	0.017	0.011
³⁄₁₆	0.18750	0.050	0.0215	0.0135

(continued)

Diameter of Stock		Thickness (T)	Depth of Angle (D)	
Fractional	Decimal		Brass	Steel
¼	0.25000	0.060	0.0255	0.016
⁵⁄₁₆	0.31250	0.070	0.030	0.019
³⁄₈	0.37500	0.080	0.034	0.021
⁷⁄₁₆	0.43750	0.090	0.038	0.024
½ to ⁹⁄₁₆	0.5000 to 0.5625	0.100	0.042	0.027
⅝ to ¾	0.6250 to 0.7500	0.120	0.051	0.032
¹³⁄₁₆ to 1	0.8125 to 1.000	0.140	0.059	0.038
1¹⁄₁₆ to 1⁵⁄₁₆	1.0625 to 1.3125	0.160	0.068	0.043
1³⁄₈ to 1⅞	1.3750 to 1.8750	0.190	0.081	0.051
2 to 2½	2.0000 to 2.5000	0.220	0.093	0.059

The thickness of the cutoff for stock 1 inch in diameter is 0.140 inch (see Table 5-3) or approximately ⅛ inch, and a circular tool of that thickness, or width, is fed 0.0035 inch per spindle revolution (see Table 5-1). Therefore, 0.574 inch ÷ 0.0035 inch = 164 spindle revolutions required for the cutoff tool.

Since 164 spindle revolutions are required for the cutoff tool and only 78 revolutions are required for the forming tool, the spindle revolutions for forming and cutoff can be overlapped. Then, as the work progresses, the spindle revolutions required for all operations that actually take time can be entered in the cam design worksheet (see Figure 5-16) in the second set of columns, eliminating those that overlap.

Spindle revolutions are also required for *indexing the turret* and *feeding stock*. To find the spindle revolutions for these operations, take a trial total of the revolutions that remain after the overlapped operations are removed.

Rough turning requires 150 revolutions, finish turning requires 150 revolutions, and cutting off requires 164 revolutions. These operations total 475 revolutions to give an approximate idea of the time required for the job.

On referring to the table for laying out cams (see Table 5-2), it can be observed that the total is within the limit of 20 seconds, the point at which the cam surfaces normally begin to govern the time required for feeding stock or indexing the turret on this machine. The revolutions required for these movements are calculated on the basis of time required for the mechanism to operate plus a few extra revolutions to allow time for setting the trip dogs. It sometimes seems, from the time trial of revolutions, that only actual time is required for feeding of stock and indexing the turret. But, on proceeding with entries in the worksheet, it is found that the cam surface governs the time required for these movements.

To feed stock or index the turret, ½ second is required (see Table 5-2). Thus, there are 25 spindle revolutions at 3000 rpm. If an additional five revolutions are added, as mentioned, 30 revolutions should be allowed for feeding stock, and another 30 revolutions for each time that the turret is indexed.

Tool clearance is also important and must be considered. It is desirable that the cross-slide tools begin to cut as soon as possible after the turret tools have completed their work. In some instances, there is no interference. However, on work that is turned to within a short distance of the chuck, ample time must be allowed for the turret tool to move backward so that it will clear the advancing cross-slide tools.

In this example, the cross-slide tools must clear a No. 22B box tool. By referring to Table 5-4, we find that $^7/_{100}$ of cam circumference for the front cross-slide tool and $^4/_{100}$ for the back cross-slide tool are required. Since there is ample time for the forming operation while the workpiece is being cut off, the clearance of $^4/_{100}$ of cam circumference for the back cross-slide tool is the only extra

Table 5-4   Approximate Clearance Between Turret Tools and Cross-Slide Circular Tools (in Hundredths of Cam Surface)

		Clearance									
		Front Cross-Slide Tool					Back Cross-Slide Tool				
No. on Tool	Turret Tools	No. 00	No. 0	No. 2	No. 4	No. 6	No. 00	No. 0	No. 2	No. 4	No. 6
20–22	Balance-turning tool		7	7				7	7		
24–26	Balance-turning tool			6	0					6	6
00A	Balance-turning tool	6					6				
00B	Balance-turning tool	6					6				

(continued)

Table 5-4   (continued)

No. on Tool	Turret Tools	Clearance									
		Front Cross-Slide Tool					Back Cross-Slide Tool				
		No. 00	No. 0	No. 2	No. 4	No. 6	No. 00	No. 0	No. 2	No. 4	No. 6
00K–20K–22	Box tool	8	7	7			6	7	6		
24–26	Box tool				7	7				6	6
00L–20L	Box tool	6	6				7	7			
00BM–20B–22B	Box tool	8	7	7			5	5	4		
00CA–20CM	Box tool	8	7				6	6			
00D–20H	Box tool	5	7				7	6			
00EA–00FA–20E–22E	Box tool	8	7	7			6	6	6		
00–00CA–11–22	Centering and facing tool	8	7	7			4	3	3		
00E–20–22	Die holder	7	6	6			6	6	6		
24–26	Die holder				6	6				6	6
00-00BA-20–00-00BA-22	Knee tool	6	5	5			4	4	4		
00–20–22	Knurl holder	7	6	6			6	6	7		
24–26	Knurl holder				6	6				5	5
00B–20B–20D–22B–22D	Pointing tool	7	7	7			5	6	6		
00C	Pointing tool	8					6				
24–26	Pointing tool				6	6				6	6
00D	Pointing tool	5					7				
24B–26B	Turret tool post				3	3					

Note—For No. 00 size machines, double the time given in table if a 3-second job.

On a 4- to 5-second job, add from four- to five-hundredths of cam surface to figures given in table for these machines.

time required to be added to the actual time for the job. The balance of the clearance, $3/100$, required for the front cross-slide clearance can be overlapped into the time required for cutting off.

After the estimate of the number of spindle revolutions required for each tool and operation has been completed and entered on the worksheet, the column headed "Spindle Revolutions After Deduction for Operations Overlapped" can be added. The sum, in this instance, is 554 spindle revolutions.

If $4/100$ of the cam circumference must be devoted to clearance, as mentioned, the balance of $96/100$ represents the total number of spindle revolutions, 554, for the operations and idle movements. Dividing the total by 96 and multiplying by 100 ($554 \div 96 \times 100$) gives a total of 576, which is the total estimated number of revolutions required to finish the workpiece.

Selecting Actual Spindle Revolutions

It can be noted that change gears are not provided for exactly 576 revolutions at a speed of 3000 rpm (see Table 5-2) so it is necessary to select either 600 revolutions or 550 revolutions because these are the nearest figures for which change gears are available. If the number of spindle revolutions is *larger* than the estimated number, *add* the extra revolutions to one or more of the cutting operations; if the number is *smaller* than the estimated number of revolutions, *subtract* from the cutting operations. Care should be taken not to attempt to subtract too many revolutions because the feeds can be increased beyond the limit for obtaining the desired finish.

In this example, the number of revolutions selected from the table is 600, which indicates that extra revolutions must be added to the total estimated revolutions (576). When the number selected is larger than the estimated total, the additional revolutions required to bring the total to 600 revolutions can be calculated as follows:

$$600 \times 0.96 = 576 \text{ revolutions}$$

Subtract   554 revolutions (96% of estimated total spindle revolutions)

      22 revolutions to be added

In this example, the 22 extra revolutions are added to the cutoff time because that is the operation to which they can be added most advantageously. It can be noted in the table (see Table 5-2) that 600 revolutions represent a 12-second operation, and the gross production per hour is 300 pieces of work.

Determining the Hundredths of Cam Surface

The hundredths portions of cam surface required for each operation and idle movement can be found by dividing the number of revolutions required by the total number of revolutions required, rounding to the nearest hundredth. The total of the hundredths portions must equal 100 or the full circumference of the cam.

Drawing the Cams

After the calculations have been completed and the portion of the cam surface required for each operation has been determined, the cam can be drawn. The usual simple and inexpensive drawing equipment is needed. In addition, a cardboard template for quickly marking the hundredths divisions on the cam circle and cam templates for drawing the quick rise and fall of the cam lever rolls when either bringing tools into position or dropping backward are needed. Many firms that have a number of machines use cam sheets printed with a cam circle divided into hundredths. The sheet also has spaces available for listing the tools used, order of operations, and other essential information. The cam design worksheet shown in Figure 5-16 is used for the workpiece shown in Figure 5-15.

The important steps in drawing the cam (Figure 5-19) are as follows:

1. Draw a long horizontal line to represent the centerline near the top of the sheet.

2. On the left-hand side, draw a vertical line to indicate the position of the face of the chuck.

3. Draw the workpiece and the cutoff and form tools as near to full scale as possible.

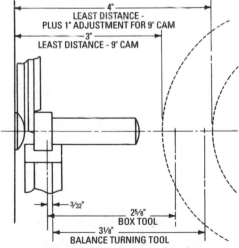

Figure 5-19 Layout diagram for the cam.

Establishing Position of the Turret Slide

The least distance of the turret from the face of the chuck on the No. 2 high-speed machine (36 spindle speeds) is 3 inches, as indicated in Table 5-5. The turret is in this position when the lead lever roll is on the circumference of a 9-inch cam. A screw adjustment on the turret slide permits setting the turret back a distance of 1 inch, if required, so that the least distance from the turret to the face of the chuck is then between 3 and 4 inches (depending on the

Table 5-5 Machine Capacities and Speeds (High-Speed, Six-Speed and Screw-Threading Machines)

	After Serial Nos. Indicated			
	Nos. OO H.S.* (3 Spds.), OO H.S.* (36 Spds.), OOG H.S.* (Beginning Serial No. 12852)	Nos. O H.S.* (3 Spds.), OO H.S.* (36 Spds.), OG H.S.* (Beginning Serial No. 8021)	Nos. 2 H.S.* (3 Spds.), 2 H.S.* (36 Spds.), 2G H.S.* (Beginning Serial No. 7352)	Nos. OO H.S.*, OOG H.S. Screw Thrd. (Beginning Serial No. 13432)
Diameter of hole through regular feed tube	25/64 in.	21/32 in.	1⅛ in.	25/64 in.
Largest stock taken in regular feeding fingers ...				
Round	⅜ in.	⅝ in.	1 in.	⅜ in.
Hex.	5/16 in.	½ in.	⅞ in.	5/16 in.
Square	¼ in.	7/16 in.	11/16 in.	¼ in.
Note—For special arrangements for greater capacities, see below.				
Greatest length that can be turned at one movement	¾ in.	1¼ in.	2 in.	
Greatest length that can be fed at one movement	1 in.	2 in.	2½ in.	1 in.
Number of spindle speeds	3 36 36	3 36 36	3 36 36	10
Fastest and slowest spindle speeds, rpm	6000 6000 6000	4220 4220 4150	3000 3000 3000	5000
	1200 200 200	850 150 155	645 100 100	1212

Table 5-5   (continued)

	After Serial Nos. Indicated			
	Nos. OO H.S.* (3 Spds.), OO H.S.* (36 Spds.), OOG H.S.* (Beginning Serial No. 12852)	Nos. O H.S.* (3 Spds.), OO H.S.* (36 Spds.), OG H.S.* (Beginning Serial No. 8021)	Nos. 2 H.S.* (3 Spds.), 2 H.S.* (36 Spds.), 2G H.S.* (Beginning Serial No. 7352)	Nos. OO H.S.*, OOG H.S. Screw Thrd. (Beginning Serial No. 13432)
Change gears give one revolution of cams				
Fastest	¾ sec.	1⅔ sec.	3 sec.	¾ sec.
Slowest	45½ sec.	176½ sec.	240 sec.	20 sec.
Actual time allowed to feed stock	¼ sec.	⅓ sec.	½ sec.	¼ sec.
Actual time allowed to index turret	¼ sec.	⅓ sec.	½ sec.	
Number of holes in turret	6	6	6	
Diameter of holes in turret	⅝ in.	¾ in.	1 in.	
Diameter of turret	3⅛ in.	4¹/₁₆ in.	5 in.	
Greatest distance tools can project from turret	2¼ in.	3¼ in.	3⅞ in. in.	
Greatest diameter of tool turret will swing	1⅝ in.	2⅝ in.	2⅞ in.	
Greatest distance between turret and chuck	2¹⁵/₁₆ in.	5⅛ in.	6¾ in.	3 in.
Least distance between turret and chuck	1⅞ in.	2½ in.	3 in.	³/₁₆ in.
Center of holes in turret to side of turret slide	⅞ in.	1⅜ in.	1½ in.	

(continued)

Table 5-5   *(continued)*

	After Serial Nos. Indicated			
	Nos. OO H.S.* **(3 Spds.),** **OO H.S.*** **(36 Spds.),** **OOG H.S.*** **(Beginning Serial No. 12852)**	**Nos. O H.S.*** **(3 Spds.),** **OO H.S.*** **(36 Spds.),** **OG H.S.*** **(Beginning Serial No. 8021)**	**Nos. 2 H.S.*** **(3 Spds.),** **2 H.S.*** **(36 Spds.),** **2G H.S.*** **(Beginning Serial No. 7352)**	**Nos. OO H.S.*,** **OOG H.S.** **Screw Thrd.** **(Beginning Serial No. 13432)**
Screw adjustments of turret slide		¾ in.	1 in.	
Rack adjustment of turret slide (one tooth)	Approx. $\frac{3}{16}$ in.			
Top of cross slide to center of spindle	1 in.	1$\frac{5}{16}$ in.	1$\frac{7}{16}$ in.	1 in.
Number of changes of speeds			20	
Die Spindle's, spindle speeds, r/min				
Fastest			7000	
Slowest			1454	
Number of cutting speeds			20	
Threading cutting speeds, rpm				
Fastest			2000	
Slowest			242	
Movement of cross slide	1 in.	1¼ in.	1¾ in.	1 in.
Distance from center of spindle to floor	46 in.	46 in.	46 in.	46 in.
H.P. required at maximum capacity	2	3	5	2¼ in.
Floor space, length	48 in. 48 in. 53 in.	62 in. 62 in. 66 in.	68 in. 68 in. 71 in.	48 in. 60 in.
Floor space, width	27 in. 27 in. 43 in.	30 in. 30 in. 47 in.	34 in. 34 in. 48 in.	27 in. 43 in.
Net weight, approx., in lb	1400 1600 1600	1850 2075 2075	2500 2875 2975	1600 1700

H.S. is high-speed steel.

distance the slide is set back). Then, if enough space still cannot be obtained, the cam must be reduced to a smaller diameter on some or all of the lobes.

To determine the distance at which to set the turret slide and the amount to reduce the lead cam lobes, refer to the dimensions of stock tools (Table 5-6). The table gives the length of the body and shank on both the No. 22 balance turning tool and the No. 22B box tool. From the table, the length of the body of the balance turning is 3⅛ (3.125) inches.

Table 5-6   Principal Dimensions of Stock Tools/Diameter

(No. 2 Machines)

Note: If two capacities are given for one tool, the capacity giving greater length and smaller diameter can be obtained only by allowing the work to pass through the hole in the shank.

	Length of Body	Length of Shank	Capacity Dia.	Capacity Length	Hole Shank	Distance from Cutting Edge to Front End of Body
No. 22DA adjustable guide; center of tongue to front of guide 3⅝ in.			Angle 30° inc.			
No. 22B adjustable guide, left-hand			Angle 30° inc.			
No. 22 angular cutting-off tool	3⁵⁄₁₆	1¾	50° to 80° inc.		⅝	
No. 22A back rest for turret	1¾	2⅜	³⁄₂₂ to 1⅛		⅝	
No. 22C back rest for swing tool	1⁷⁄₁₆	2¼	½		½	
No. 22 balance turning tool	3⅛	1¾	¼ to 1	2⅝	½	
No. 22C balance turning tool	3	1¾	¼ to 1	2⅝	½	
No. 22A box tool	2⅝	2¼	⅝	2	½	
No. 22B box tool	2⅝	2¼	³⁄₁₆ to ⅝	2	½	³⁄₃₂

(continued)

Table 5-6   *(continued)*

(No. 2 Machines)

Note: If two capacities are given for one tool, the capacity giving greater length and smaller diameter can be obtained only by allowing the work to pass through the hole in the shank.

	Length of Body	Length of Shank	Capacity		Hole Shank	Distance from Cutting Edge to Front End of Body
			Dia.	Length		
No. 22G box tool, three blades, adjustable; greatest distance between blades 1⅞ in.; least $^{11}/_{16}$	3	2	$^7/_8$	2⅜	½	⅛
No. 22E box tool, roller back rest	3⅛	2	$^5/_{16}$ to $^5/_8$	3	$^{11}/_{16}$	⅛
			$^5/_8$ to $^3/_4$	2⅝	$^{11}/_{16}$	⅛
No. 22FA box tool, roller back rest, left-hand	3⅛	2	$^5/_{16}$ to $^5/_8$	3	$^{11}/_{16}$	⅛
			$^5/_{16}$ to $^7/_8$	2⅜		
No. 22D centering and facing tool, center drill $^5/_8$ in.	1¾	2¾			$^5/_8$	$^1/_{16}$
No. 22CA centering and facing tool, center drill $^5/_8$ in.	1¾	2¾			$^5/_8$	$^1/_{16}$
No. 22 combination drilling and tapping attachment						
Drill	1½		⅛			
Tap	$1^9/_{16}$		Brass $^3/_8$–24			
			Steel $^1/_4$–32			
No. 22 cross drilling attachment, center of tongue to end of drill spindle $2^{15}/_{16}$ in.			¼			
No. 22B die holder, releasing, extreme pull out ⅛ in.	2⅜	2	½	2⅜	¼	

Table 5-6   *(continued)*

(No. 2 Machines)

Note: *If two capacities are given for one tool, the capacity giving greater length and smaller diameter can be obtained only by allowing the work to pass through the hole in the shank.*

	Length of Body	Length of Shank	Capacity		Hole Shank	Distance from Cutting Edge to Front End of Body
			Dia.	Length		
No. 22 die holder, extreme pull out ½ in.	2⅜	2	½	1¾	⅜	
No. 733B die holder, acorn, releasing	3¼	2	⅜–½ ½–⅝	3⅛ 2⅛		
No. 22 drill holder	1½	1¾			⅝	
No. 22 drilling attachment	1½		⅜		¼	
No. 22E fixed guide; center of tongue to front of guide 3½ in.						
No. 22D fixed guide; center of tongue to front of guide 3⅝ in. (For L.H. swing tools)						
No. 22 floating holder	1⁹⁄₁₆	1¾			¹¹⁄₁₆	
No. 22A hollow mill, plain, held in floating holder	2⅝	1¾	⁷⁄₁₆			
No. 22C			¹¹⁄₁₆			
No. 22 knee tool	3	2	²¹⁄₃₂ 1⁷⁄₁₆	3 1¹¹⁄₁₆	¹¹⁄₁₆	⅛
No. 22B knee tool	3		²¹⁄₃₂ 1⁷⁄₁₆	3 1¼	¹¹⁄₁₆	⅛
No. 22 knurl holder for turret	2⅜	2³⁄₁₆	³⁄₁₆ to ¹⁵⁄₁₆	1⅞	½	

(continued)

Table 5-6   *(continued)*

(No. 2 Machines)

Note: If two capacities are given for one tool, the capacity giving greater length and smaller diameter can be obtained only by allowing the work to pass through the hole in the shank.

	Length of Body	Length of Shank	Capacity		Hole Shank	Distance from Cutting Edge to Front End of Body
			Dia.	Length		
			$^3/_{16}$ to $1^1/_8$	$^1/_2$		
No. 22AA knurl holder, side; center of tool to center of knurl $1^7/_{32}$ in.			Width $^1/_4$			
No. 22BA knurl holder, top, back; center of circular tool to center of knurl $2^3/_{16}$ in.			$1^1/_8$ width $^1/_4$			
No. 22C knurl holder, bottom; center of circular tool to center of knurl, $2^3/_8$ in.			$1^1/_4$ width $^1/_4$			
No. 22A pointing tool holder, for circular tools	$2^3/_8$	$1^{25}/_{32}$				
Nos. 22B and 22D pointing tool, box stock stops, length $2^1/_2$ in., $2^7/_8$ in., $3^1/_4$ in., $3^5/_8$ in., 4 in.	2	$2^1/_2$	$^7/_8$		$^5/_8$	$^1/_2$
No. 22M recessing tool, diameter of shank 1 in.	2	$1^5/_8$				
No. 22BA swing tool, actual projection of body $2^3/_4$ in.	$2^{17}/_{32}$	2	$2^1/_{32}$ 1	3 $1^3/_4$	$^{11}/_{16}$	

Table 5-6   (continued)

(No. 2 Machines)

Note: If two capacities are given for one tool, the capacity giving greater length and smaller diameter can be obtained only by allowing the work to pass through the hole in the shank.

	Length of Body	Length of Shank	Capacity Dia.	Capacity Length	Hole Shank	Distance from Cutting Edge to Front End of Body
No. 22CA Swing Tool, actual projection of body $2\frac{1}{16}$ in.	$1\frac{7}{8}$	2	$\frac{21}{32}$ 1	3 $1\frac{1}{8}$	$\frac{11}{16}$	
No. 22HA swing tool, recessing; swing from center $\frac{1}{8}$ in.	$2\frac{1}{4}$	2			$\frac{5}{8}$	
No. 22KA swing tool, knurling or thread rolling	$2\frac{3}{16}$	$2\frac{1}{2}$	$\frac{21}{32}$ $1\frac{3}{8}$	3 $1\frac{1}{4}$	$\frac{11}{16}$	$\frac{1}{4}$
No. 22LA swing tool, actual projection of body $2\frac{3}{4}$ in.	$2\frac{15}{32}$	$2\frac{1}{2}$	$\frac{21}{32}$ 1	3 $1\frac{3}{4}$	$\frac{11}{16}$	
No. 22 tap holder; extreme pull out $\frac{1}{2}$ in.	$1\frac{5}{8}$	2				
No. 22B tap holder, releasing; extreme pull out $\frac{1}{8}$ in.	$2\frac{1}{16}$	2				
No. 22 tapping attachment; extreme pull out $\frac{5}{16}$ in.	$1\frac{9}{16}$		Brass $\frac{3}{8}$;—24 Steel $\frac{1}{4}$–32			

Lay out the distance (3.125 inches) on the scale drawing (see Figure 5-19), beginning at the point where the tool finishes its cut at the shoulder of the piece and extending backward toward the turret position. Repeat this procedure with the box tool. Both these points are established beyond the position of the turret when it is at its shortest distance from the chuck.

If the full adjustment of the turret slide is used, there will be barely enough space for the balance-turning tool. On a 9-inch cam blank, the shortest distance from the turret to the chuck is 3 inches. Then, the additional distance desired for the balance-turning tool can be obtained by adjusting the turret slide backward 0.5 inch plus a 0.5-inch reduction on the cam.

Note
With a 9-inch cam, a length of 2 inches can be turned on the workpiece. With an 8-inch cam, a 1.5-inch length can be turned. If none of the throws on the cam exceed 1.5 inches and reductions are not required, an 8-inch cam can be used. By using special long turret travel parts, 3 inches can be turned with a 9-inch cam.

At all times, it is desirable to have the tools set back into the turret, as close to the turret as possible. If the shanks extend through the holes into the center of the turret, be careful that two tools in succession do not extend through since the shanks can interfere with the operation of the turret on the machine.

Drawing the Cam Lobes
First, draw a 9-inch circle, and divide the circumference into hundredths using a template. From the cam design sheet (see Figure 5-16), indicate the hundredths divisions or graduations on the cam circle as to where operations are to begin and end for the turret slide or lead cam.

The hundredth division or the graduation line at the top, directly above the hole for the locating pin, is the "zero" line. The stop for the stock is in position for the first five divisions of the circle, and a "dwell" is drawn on the cam surface. If this is drawn to the full height of the cam, a high narrow lobe results. Therefore, it is better to reduce the cam at a point 1.25 inches from full height. A stop of proper length can be used in the turret to compensate for this reduction.

From the 5th to the 10th division marks, the turret is double indexed and dropped backward to the required position for the beginning the first turning operation. On the 10th division radius line, measure downward and locate a point 1.5 inches from the outside diameter plus a 0.5-inch reduction; this represents the beginning of the balance-turning tool lobe.

Locate the cam template at the center of the drawing with an ordinary pin through the hole in the template (Figure 5-20). Place the arm that has a drop marked "6 to 40 seconds" so that the edge is on the 5th division line, and follow the edge with a pencil. Set the compass to the radius of the cam roll (⁵⁄₁₆ or 0.3125 inch), and connect the line of drop with the point on the 10th division radius line where the lobe for the balance-turning tool begins. The space for indexing allows some clearance after the roll has ended contact with the line of drop and before it starts on the balance-turning lobe. This clearance should be provided for because it is not always safe to drop from one lobe directly to the starting point of the next lobe.

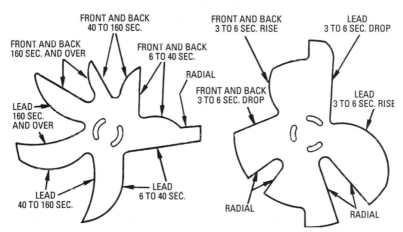

Figure 5-20 Two cam templates for Brown & Sharpe No. 2 machines. One is used for those jobs that are within the range of "6 to 240 seconds" per piece (left), and the other is a high-speed template for jobs within the range of "3 to 6 seconds" per piece (right).

How to Develop a Cam Lobe Curve

The balance-turning tool is in operation from the *10th to the 35th division line*. To develop a uniform curve with constant rise, divide the throw into any number of equal parts, and then divide the arc of the cam circumference so that it occupies into the same number of equal parts. Swing the arc at the division points of the throw, and draw the radii to intersect these arcs from each division point on the circumference of the cam. Use a scroll to draw a spiral curve through the points of intersection to give the outline of the cam lobe.

The indexing space from the *35th to the 40th division line* and the lobe for the box tool from the *40th to the 65th division line* are drawn in exactly the same manner as has been described previously. Then, the outline drops back to the full depth of the throw, withdrawing the turret slide while the cross slides are in operation. The turret slide advances again, according to the rise of the template, in time for the stock to stop again at the correct distance (zero).

The cross-slide tools begin to cut as soon as possible after the box tool is finished. The diameter of the cross-slide cams is 7 inches. Use the same center, but indicate the outline of the front cross-slide cam by dots and the back cross-slide cam by alternate dots and dashes.

As indicated on the worksheet, a $4/100$ space is required for clearance between the box tool and the back cross-slide tool. So, the cross-slide tool begins at the *69th division line and ends at the "zero" division line* because the operation requires $31/100$ of the cam surface.

The starting point of the throw is measured from outside the cam, and the rise from the full depth of the cam to the starting point of the lobe is made by the cross-slide curve on the cam template. The lobe curve is developed in the same manner as described previously. There is no difference in drawing the lobe for the front slide or forming tool except that more clearance is necessary, and, during the last hundredth space (from $84\frac{1}{2}$ to 85), the curve is on the true circumference of the cam so that the tool dwells for several revolutions before dropping back.

Making the Drawing

You should place all essential information on a drawing for the purpose of making a record and make the job easier for the person making the cams. Note that a dimensioned drawing can be made in the blank space on the cam design worksheet. At the top of the drawing, the distance to adjust the turret from the chuck (with the cam roll on top of the stop lobe) can be indicated for the convenience of the operator setting up the job.

It is important for the toolmaker to understand the design of such special purpose cams. However, industry is adapting new technology to the task of designing such cams. A computer program for the type of special purpose cam needed is prepared. The necessary information is fed into a computer connected to a plotter. The cam is drawn by the plotter. After the program is prepared, the process takes less than 15 minutes total time. Manual methods usually took from 25 to 30 hours for the same task. With such a saving of time, it should be expected that many more tasks will be assigned to the computer.

How to Machine Cams

Any good homogeneous sheet steel of the required grade of steel is suitable for cams. When they are hardened properly, cams from such stock can withstand the demands of all ordinary requirements.

Generally, it is more convenient to make the cams from stock blanks because they can be turned to correct size and drilled for the shaft and locating pin, and the cam surface can be graduated into hundredths division lines to facilitate laying out the outline. If the blanks are made in the shop, the holes must be of the correct size so that there will be no play of the cam on the shaft.

Transferring the Cam Outline

To get successful results, the cam outline as plotted on the drawing should be carefully and closely reproduced on the blank. A quick and handy method of transferring the outline is to place a piece of carbon paper and a piece of stiff plain paper beneath the drawing and trace the outline of the drawing with a dull steel point, which makes a carbon outline on the plain paper. This can be cut out and used as a template for scribing the outline on the cam blank.

Machining the Cam Outline

After the outline is laid out on the cam blank, it is necessary first to remove the excess metal close to the outline between the lobes. This metal can be removed by drilling a row of holes near the outline, breaking off the excess metal; a band saw or similar means can be used.

Three methods of finishing the curves of cam lobes are in use. The simplest method is filing the excess metal down to the outline. This is a slow procedure and requires considerable care to obtain a smooth curve. Cam lobes produced by this method can be satisfactory if the limits of accuracy are not too close, but, on finer classes of work, the outline of the cam lobe must be an accurately increased curve for the tool to be fed uniformly throughout the entire throw of the lobe. An extremely accurate curve can be produced on the milling machine geared for the correct lead.

A second method of finishing the cam lobes after they have been drilled and broken off to the approximate outline requires the use of a universal or plain milling machine equipped with an index head (Figure 5-21). The cam blank is mounted on the index head spindle, which should be parallel to and directly beneath the

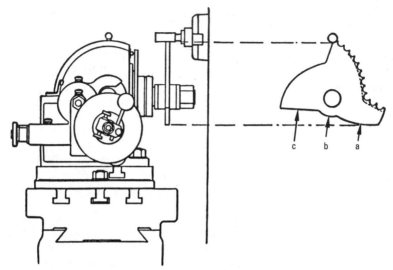

Figure 5-21 Milling machine setup for finishing the cam lobes.

machine spindle. An end mill that has the same diameter as the cam roll is used in the setup. As shown in the setup, the end mill is set at the top of the lobe and then indexed a small amount. Then, the table is fed a vertical distance corresponding to the rise of the lobe in the distance that the blank is indexed. These operations are continued until the curve of the lobe is completed. The surface that is produced may require a slight amount of filing, depending on the fineness of the division selected for indexing the blank and the finish desired.

The third method of finishing the cam lobes is nearly as simple and is much preferred to the other two methods since it produces a true uniform curve without recourse to filing. Either a plain or a universal milling machine equipped with a spiral headstock and a vertical milling attachment can be used.

The spiral head is geared to the table feed screw (as in cutting spirals), and the cam blank is fastened to the end of the index spindle.

An end mill is used in the vertical milling attachment, which is set to mill the periphery of the cam at right angles to its sides. In other words, the axes of the spiral head spindle and the attachment spindle must always be parallel in this method of milling cams. The cutting is done by the teeth on the periphery of the end mill.

To illustrate the basic principle of this method, the spiral head is elevated to 90°, or at right angles, to the surface of the table and is geared for any given lead. As the table advances and the blank is turned, the distance between the axes of the index spindle and the attachment becomes less, which indicates that the cut becomes deeper and the radius of the cam is shortened, producing a spiral lobe having a lead that is the same as that for which the machine is geared (Figure 5-22).

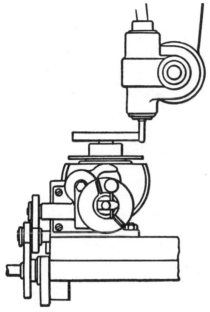

Figure 5-22 Setup for machining a cam blank with the axis of the spiral head at 90° to the surface of the table.

If the same gearing is retained and the spiral head is set at "zero" reading, or parallel to the surface of the table, it is apparent that the axes of the index spindle and the attachment spindle are parallel (Figure 5-23). Therefore, as the table advances and the blank is turned, the distance between the axes of the index spindle and the attachment spindle remains the same. As a result, the periphery of the milled blank is concentric, or the lead is "zero."

If the spiral head is elevated to any angle between 0° and 90° (Figure 5-24), the amount of lead given the cam will be between the lead for which the machine is geared and "zero." Hence, cams with a large range of different leads can be obtained with a single set of change gears. The problem of milling the lobes of a cam is reduced to determining the proper angle to set the head to obtain a given lead. Manufacturers can provide a set of tables from which the proper angles and change gears can be selected.

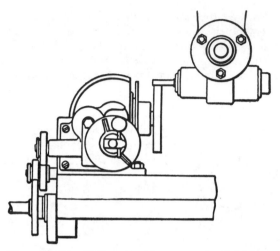

Figure 5-23 Milling machine setup for finishing cam lobes in which the axis of the spiral head is parallel to the surface of the table.

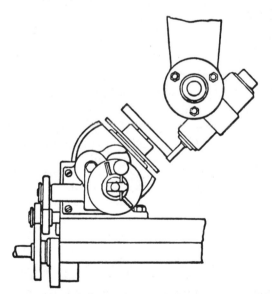

Figure 5-24 Setup for machining a cam with the spiral head at an oblique angle to the surface of the table.

Summary

The cam is a device used to convert rotary motion into reciprocating motion. Cams can be generally divided into two classes: uniform motion and uniformly accelerated cams. A cam is used to convert rotary motion into varying-speed reciprocating motion. This basic action results when an irregularly shaped disc revolves on a shaft and imparts irregular motion to a follower. Cam followers have a variety of shapes.

The motion of a cam is transmitted to a follower, which remains in contact with the acting surface as the cam revolves. The uniform motion cam moves the follower at the same rate of speed throughout the stroke from the beginning to the end. This means that if the movement is rapid, there is a distinct shock at the beginning and end of the stroke because the movement is accelerated from zero to the top speed of the uniform motion almost instantly and is stopped just as abruptly.

The uniformly accelerated motion cam is one in which the speed is slow at the beginning of the stroke, accelerated at a uniform rate until maximum speed is attained, and then uniformly retarded until the follower is stopped (or nearly stopped) when the reversal takes place.

Review Questions

1. What is a cam?
2. What are the two basic types of cams?
3. Name four common shapes for cam followers.
4. What is harmonic motion?
5. What is gravity motion?
6. Describe tangential cam action.
7. What are displacement diagrams?
8. What is valve lift?
9. The _____ with which the valves in a gasoline engine open and close is determined by the slope of the valve opening and closing faces on the cam.
10. The design of cams for exhaust valves is different from that of _____ valves.
11. A _____ can be used to assemble data for making a cam.
12. Lever _____ can be used to calculate close timing in cam design.

13. What is meant by tool clearance?

14. What is a cam lobe?

15. After the outline is laid out on the cam blank, it is necessary first to remove the excess metal close to the outline between the _____.

Chapter 6

Dies and Diemaking

A *die* is a tool intended to *cut* or *shape* sheet metal or similar materials. Basically, the die proper consists of the die block and the punch.

The *die block* is the part of the die that contains a hole that has the same outline as the piece that is to be punched. The *punch* fits into the hole in the die block. It is the part to which power is applied for performing the cutting operation. The hardened and tempered die block does the actual cutting. The *stripper plate* in the die block strips the stock from the punch on the releasing stroke. The *guide strip* merely guides the stock, and the *gage pin* is used to gage the location of the holes punched in the stock.

In some die designs, the parts of the punch are integral in one piece, but they are separate pieces united by suitable means in other designs. The *shank* attaches the punch to the ram of the press, and the *collar* of the *punching block* takes the thrust.

Figure 6-1 shows typical sheet metal stampings produced for the electronics industry.

Cutting or Punching Dies

Many types of dies are required to handle the large variety of cutting and shaping operations encountered in general practice. Dies can be classified with respect to the basic operation that they perform:

- Cutting dies
- Shaping dies
- Combined cutting and shaping dies

The cutting dies are the simplest form of dies. The part that is cut out can have any shape, depending on the shape of the punch.

Plain Die

The *plain* (or *blanking*) die is the most frequently used type of die. Depending on the shape of the punch, the *blank* can have any shape. This die is used to punch a given shape from sheet metal in one continuous cutting operation (Figure 6-2). The material can be either stock or a blank obtained from a preceding operation. Sometimes the blank punched from the stock is the part to be used,

Figure 6-1 Typical sheet metal stampings. *(Courtesy Paul and Beekman, Inc.)*

or it can be only waste material, and the punched stock is the portion that is to be used.

Figure 6-3 shows a typical sequence of operations of a plain die in punching sheet metal. A strip of sheet metal or stock is placed between the die and the stripper plate. On the downward stroke, the punch "punches out" the blank from the stock, and the stock is held down by the stripper plate to prevent lifting on the upward stroke of the punch. If several punches are combined for blanking several pieces simultaneously, they are called *multiple dies*.

Self-Centering Die

A *self-centering die* is provided with a small conical point (Figure 6-4) so that it does not require a gage pin for locating the holes punched in the stock. The self-centering die is used chiefly for punching holes. The stock must be center-punched at the points where the holes are required and then placed under the punch so that the conical point registers with the center punch mark on the stock.

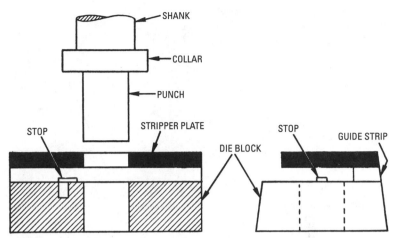

Figure 6-2 Diagram showing basic construction of a plain die.
(Courtesy Paul and Beekman, Inc.)

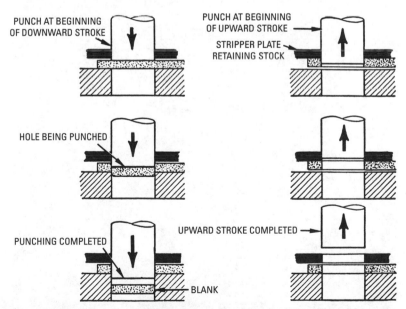

Figure 6-3 Sequence of operations of a plain die in punching sheet metal, showing the downward (or punching) stroke (left) and the upward (or releasing) stroke (right).

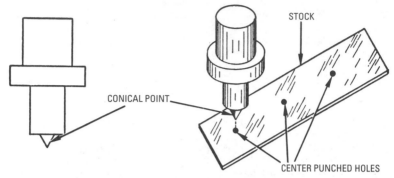

Figure 6-4 A self-centering die (left), and view showing the die in correct position (right).

Figure 6-5 Parts that require shaping. *(Courtesy Paul and Beekman, Inc.)*

Shaping Dies

Shaping dies are used to change the shape of the stock. They do no cutting. The term *punch*, strictly speaking, is a misnomer because it does not operate as a punch in shaping operations. The punch is used to press the stock against the die to bend the stock to the desired shape. Figure 6-5 shows some sheet metal parts that require shaping. As will be noted, some of these are combined with cutting or punching. These may be separate operations or combined, depending upon the particular job.

Plain Bending Die

A typical die used to bend stock to the desired shape is the *plain bending die*, shown in Figure 6-6. The contour of the punch and die determines the final shape of the stock.

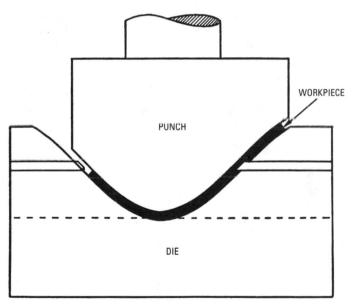

Figure 6-6 A plain bending die.

The springiness of the stock or material is an important factor in this type of die operation. In bending stock to irregular shapes (Figure 6-7), an allowance must be made for this factor in the shape of the die so that the work is bent slightly beyond the required angle or shape. This allowance is necessary to compensate for the backward spring of the stock when it is released from the die.

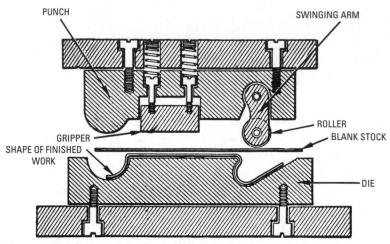

PUNCH

SWINGING ARM

ROLLER

BLANK STOCK

GRIPPER

SHAPE OF FINISHED
WORK

DIE

Figure 6-7 A typical die for bending stock to an irregular shape.

Curling Die

A *curling die* is a special form of bending die in which part of the stock is bent to form circular beads around the edge of drawn cylindrical parts (Figure 6-8). Figure 6-9 shows the progressive operations in curling a bead around a cylindrical vessel.

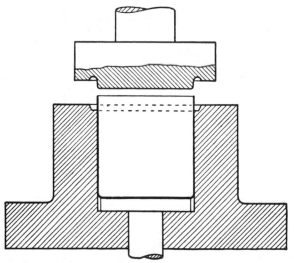

Figure 6-8 A curling die.

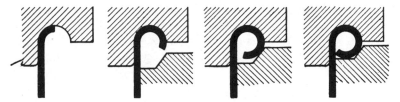

Figure 6-9   Progressive operations in curling a bead around a cylindrical vessel.

The curling punch forces the metal outward to increase the circumference at the edge of the metal. The metal will crack if it is overstretched. The stretching action of the metal increases as the diameter of the bead increases. Ordinarily, the maximum diameter of the bead that can be formed is ³⁄₁₆ inch for tinplate and sheet iron. An extra large bead can be obtained without cracking the metal by annealing the metal.

Wiring Die

The *wiring die* is a modification of the curling die in which provision is made for curling the edge of the workpiece over a wire. Figure 6-10 shows the basic diagram of a wiring die. The shape of the punch is the same as for the curling die, but the die proper (floating ring) is constructed to "float" on springs, as shown in the diagram.

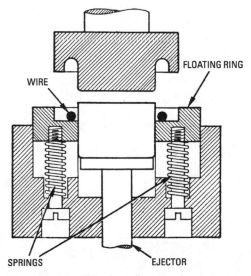

Figure 6-10   Diagram showing basic parts of a wiring die.

In preparation for the wiring operation, a ring of wire is placed in position around the projecting edge of the workpiece. As illustrated in the progressive operations (Figure 6-11), the punch, as it descends, coils the edge of the workpiece around the wire in successive steps.

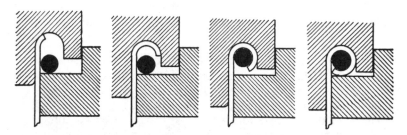

Figure 6-11   Progressive operations of a wiring die.

Bulging Die

The basic principle of operation of the bulging die depends on the expansive action of a springy material placed under pressure. A rubber disc is subjected to pressure by a descending plunger (Figure 6-12).

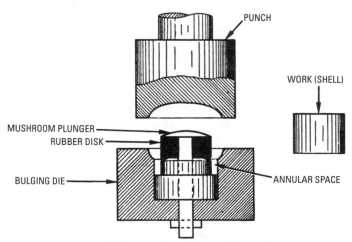

Figure 6-12   Diagram of basic parts of a bulging die (left) and a section of the workpiece (right).

In typical operations, the workpiece consisting of a shell drawn upright (see Figure 6-12) is placed directly above the mushroom plunger, as shown in Figure 6-13. As the plunger descends, the rubber disc is expanded outward to expand or "bulge" the shell into the curved chamber formed by the punch and bulging die. A cylindrical projection is formed as the work enters and traverses the annular space.

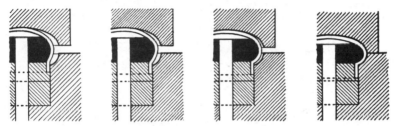

Figure 6-13   Progressive operations of a bulging die.

After the progressive die operations have been completed, the workpiece appears as shown in Figure 6-14. Similar types of work, such as covers or lids for some kinds of cans or containers, can be formed on the bulging die.

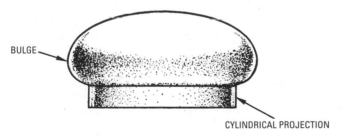

Figure 6-14   Appearance of a typical workpiece that has been formed on a bulging die.

Combination Punching and Shaping Dies

Various combinations of punches and dies can be used to perform both the cutting and shaping operations. This is illustrated in the double-action die shown in Figure 6-15.

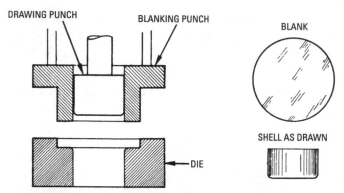

Figure 6-15   A double-action die (left). Also shows the workpiece after blanking and drawing (right).

Double-Action Dies

Two concentric punches are used in this type of *double-action die.* An outer punch is used to perform the operations of cutting and holding the blank. The inner punch is a forming or drawing punch.

In actual operation (see Figure 6-15), the cutting or blanking punch first descends to cut a blank. When the blank has been cut, the punch forces the blank onto the top of the die and presses it against the annular ring with enough pressure to prevent wrinkling. Then, it is allowed to recede into the forming part of the die during the drawing process (Figure 6-16). Then, the inner or drawing punch descends to perform the drawing operation and finish the shell.

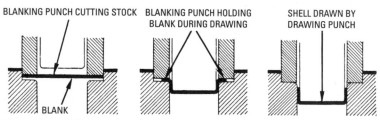

Figure 6-16   Sequence of progressive operations of the double-action blanking and drawing die.

The drawing dies usually employ a punch to shape or draw the work, and, in some instances, a second drawing operation (or redrawing) is performed on work from the drawing die. Figure 6-17 shows some typical drawn parts required by the electronics industry.

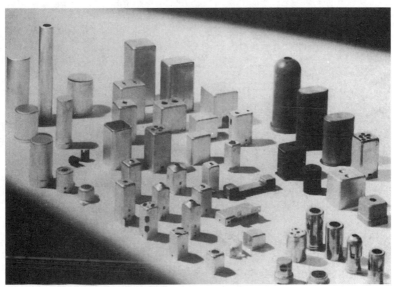

Figure 6-17   Some typical drawn parts for the electronics industry.
(Courtesy Paul and Beekman, Inc.)

Plain Drawing Die

The *plain drawing die* is intended only for shallow drawing because there is no pressure holder on the blank to prevent its wrinkling. Figure 6-18 shows a drawing punch with a diameter equal to the diameter of the hole minus twice the thickness of the blank or stock. The die hole is tapered slightly.

The blank fits in the recess at the top of the die. After the drawing operation is completed, the bottom edge of the die strips the shell from the punch on the upward stroke.

Redrawing Die

The *redrawing die* performs a second operation on a workpiece that comes from the drawing die. A second drawing operation is performed, which reduces the diameter of the metal and increases its length. In other words, a cup or shell already formed from the stock is redrawn.

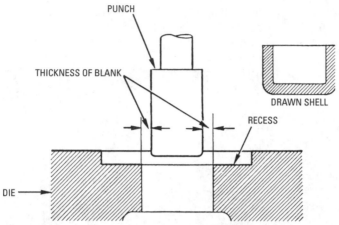

Figure 6-18 A plain drawing die (left) and completed section of work (right).

The redrawing die is similar to the drawing die in that the recess at the top of the die fits the shell that is to be redrawn instead of being made to fit a blank that is to be drawn. Figure 6-19 shows a redrawing die with the shell in position for the redrawing operation.

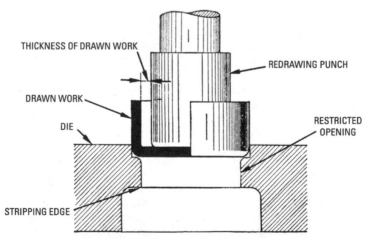

Figure 6-19 A redrawing punch.

Figure 6-20 shows successive steps in the redrawing operation. During the downward stroke of the punch, the drawn shell is forced through the restricted opening. This squeezes the shell to a smaller diameter.

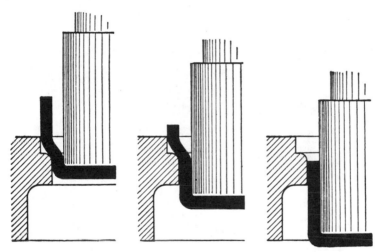

Figure 6-20   Successive steps in the operation of the redrawing punch.

As the perimeter is reduced, the excess metal in the shell is squeezed lengthwise. Therefore, the shell is lengthened by the redrawing operation. After the redrawing operation is completed, the workpiece is stripped or removed from the punch by the stripping edge.

Redrawing dies are sometimes called *reducing dies* because the redrawing operation reduces the diameter of the work. If considerable elongation is desired, the work is passed through a number of redrawing dies.

Gang and Follow Dies

A *gang die* is one in which two or more punches are combined in a single punching head. As many holes as there are mounted, punches can be punched simultaneously in a single operation (that is, on the downward stroke of the assembly). Figure 6-21 shows a diagram of the basic parts of a gang die. The electronic chassis shown in Figure 6-22 is an example of work done with a gang die.

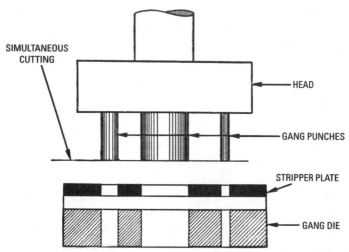

Figure 6-21 A gang die. Note that all punches operate at the same time.

A *follow die* is a modified gang die in which several successive operations are performed progressively. The difference between the gang die and the follow die is illustrated in Figure 6-23.

Compound Die

Figure 6-24 shows the basic construction of a compound die. This type of die is constructed with a die in the upper punch and a punch in the lower die. That is, it contains two punches and two dies. By the combined action of the two punches and the two dies, blanks are formed to the desired shape for which the die was designed.

Compound dies are different from plain and follow dies in that the basic punch and die elements are combined so that both the upper and lower members contain the equivalent of a punch and die, as well as stripper plates or ejectors.

In actual die operation (Figure 6-25), the upper die descends and depresses the stripper plate. As the downward movement is continued, the blank is cut from the stock by the telescoping action of the upper and lower dies, the lower die acting as a punch. Both dies function as punches as well as dies (that is, the upper die acts as an external punch, and the lower die acts as an internal punch).

At the same time that the blank is cut, the punch pierces a central hole in the upper element. As the ram ascends, the blank is forced out of the upper die by the ejector, and the stripper plate pushes

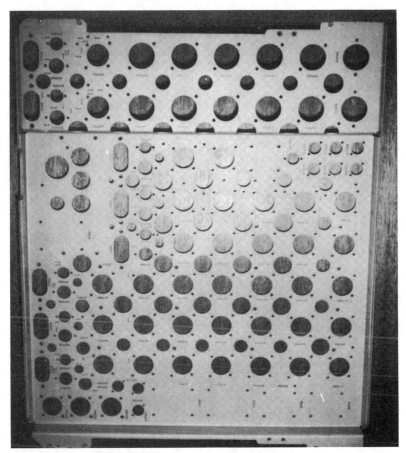

Figure 6-22 An electronic chassis with a variety of punched holes.
(Courtesy Paul and Beekman, Inc.)

the work off the lower die. The metal punched from the hole falls through the opening in the lower die.

Delicate parts can be made with a compound die. Gear punchings for meters, clocks, and so on, are typical of the accurate work that can be produced quickly and in large quantities with compound dies.

Miscellaneous Dies
Numerous dies are made for various purposes. In addition to the dies already described, numerous other dies should be mentioned.

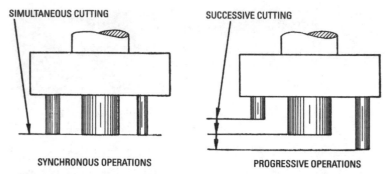

Figure 6-23 Note the difference between a gang die (left) and a follow die (right). Note the progressive cutting operations of the follow die.

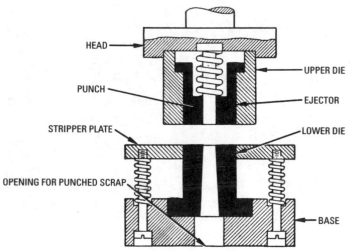

Figure 6-24 Diagram showing basic construction of a compound die.

A *burnishing die* is constructed slightly smaller at the bottom than at the top. Since the bottom of the die is a little smaller than the top, the work receives a high finish as it is forced downward through the die.

A *combination die* can be used to perform a combination of operations in a single stroke. The blank can be cut. As the stroke progresses, the edge can be turned down, and the piece drawn into the required shape.

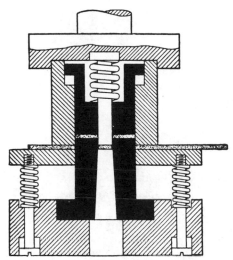

Figure 6-25   Diagram of a compound die, showing the position at the end of the stroke.

A *trimming die* is used to remove excess metal. It is designed so that the excess metal is removed from the edges or ends of various drawn and formed workpieces.

The *fluid dies* are designed for forming fancy hollowware from soft metals by means of hydraulic pressure. The shell that is to be formed is filled with liquid and enclosed within the die. Then, by means of a plunger, hydraulic pressure can be applied to force the shell outward against the contour of the enclosing walls of the die.

A *perforating die* is a multiple die. Many small-diameter punches can be grouped together to form the multiple die.

A *reducing die* is a type of redrawing die. It is designed to reduce only a portion of the shell instead of the entire length, as in regular redrawing dies.

Three independent movements can be made by the *triple-action die*. It can be used to perform such operations as cutting or blanking, drawing and stamping, or embossing.

Frequently, a part cannot be made in a single operation but requires several. Often, these are made in such a way that the material is moved through a series of operations. These are known as *progressive dies*. Figure 6-26 shows a 14-stage progressive die that forms the automobile part in the foreground. The workpiece in the center shows the various stages from the flat strip at the bottom to

Figure 6-26 A 14-stage progressive die. *(Courtesy Paul and Beekman, Inc.)*

the finished piece at the top. The work is moved up one stage automatically with each stroke of the press.

Diemaking Operations

The choice of a suitable lubricant, selection of the proper material, and layout of the die are all important operations that must be considered in diemaking.

Lubricants

Although some shops do not use a lubricant for working sheet metal, a lubricant is recommended for best results. A heavy animal oil can be used for cutting steel and brass. Lard oil can be used, but it is more expensive. A mixture of equal parts of oil and black lead can be used for drawing steel shells.

When working with German silver or copper stock, a thin coat of sperm oil can be used. Lard oil can also be used on these materials.

The lubricant can be spread over the stock by coating a single sheet heavily with the lubricant and passing it through a pair of rolls. Then, several sheets can be lubricated lightly by passing them through the rolls before the supply of lubricant on the rolls is exhausted. This is also a good method for lubricating work that is to be drawn because the thin coating of oil will disappear during the drawing operation, and cleaning the shells is unnecessary after the operation has been completed. A brush or pad is objectionable for applying lubricant because the coating of oil is usually too thick and requires cleaning after the operation has been finished.

Clean soap and water is recommended as a lubricant for drawing brass and copper. For aluminum stock, lard oil, tallow, or Vaseline is recommended.

Materials for Making Dies

In most instances, a good grade of tool steel is used for making punches and dies. The steel should be free of harmful impurities. Sometimes, the body of the die can be made of cast iron with an inserted steel bushing to reduce the cost of material. An advantage of this type of construction is that an insert can be replaced when it becomes worn. Soft steel that has been case-hardened does not change its form as readily as tool steel, and any minute changes in form can be corrected readily because the interior is soft.

Internal strains or stresses are set up in steel during the manufacturing process. In diemaking operations, these stresses must be relieved before the die is brought to its final size, or they will cause distortion. The presence of stresses cannot be determined in the steel beforehand, but the diemaker can relieve the stresses in the steel by annealing after the die has been roughed out.

Laying Out Dies

Before laying out a die, the diemaker must determine the most economical way to punch the stock so that there will be a minimum waste of metal. This determines the layout for the die and the location of the gage or stop pin.

The simplest method of determining the layout is to cut several pieces to fit the outline of the blanks to be made. Then, the pieces can be arranged in various ways to determine the most economical arrangement for punching the stock. If the problem has been solved correctly, the maximum number of blanks can be obtained from a given unit of stock.

Determining Minimum Waste

The cut-and-trial method can be used to determine the best layout for the diamond-shaped pattern (Figure 6-27). Paper templates can be cut out and arranged on the sheet of stock, as shown in Figure 6-28.

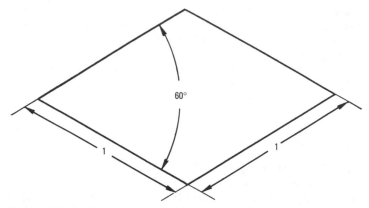

Figure 6-27 Diamond-shaped pattern, used to illustrate a cut-and-trial method of determining the way to cut the stock for minimum waste of material.

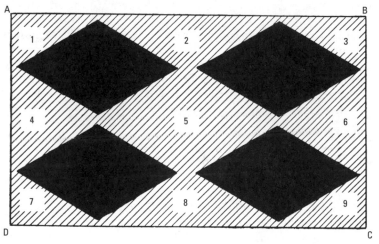

Figure 6-28 Trial layout for the diamond-shaped pattern.

Draw the enclosing rectangle *ABCD* to represent the amount of stock required to make four diamond-shaped blanks.

As shown in Figure 6-28, areas 1, 2, and so on, represent considerable waste metal or scrap. Then, rearrange the templates (Figure 6-29) and lay out another rectangle. Note that the rectangle *ABCD* is reduced to the rectangle *abC'D*. The saving in stock is represented by the rectangles *AB'ba* and *B'BCC'*, leaving considerably less scrap metal when compared to the scrap metal left in areas 1, 2, and so on, in Figure 6-28. In this instance, the difference in waste can be determined by eye, but this is not always possible.

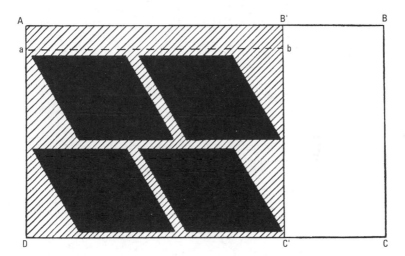

Figure 6-29 A second trial layout for the diamond-shaped pattern. This is the correct layout for minimum waste of material.

Locating the Gage Pin

After positioning the workpiece for minimum waste of stock, the location of the gage pin should be determined. The gage pin should be located so that it will hold the workpiece in position without shifting when the stock is pushed against it.

A typical pattern (Figure 6-30) can be used to illustrate a method of determining the location of the gage pin. In the illustration, the pattern is in the correct position for minimum waste of material.

The pin can be located with respect to the hole in the design or with respect to some other portion of its contour. As shown in Figure 6-31, lay off the lines *ab* and *cd* to indicate the edges of the stock, and place the pattern in position with point B as the center of the hole. Draw the axis *XX'* through point B.

The position of the part of the stock that forms the stop determines the amount of stock that remains between the punched holes. The distance between one position of the stock and the gage pin can be determined from the patterns laid on the stock. To determine

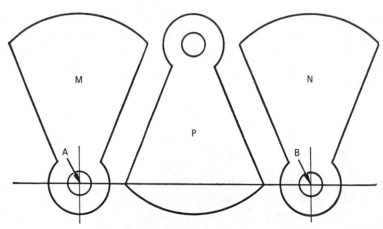

Figure 6-30 A typical pattern for showing a method of determining the location of the gage pin.

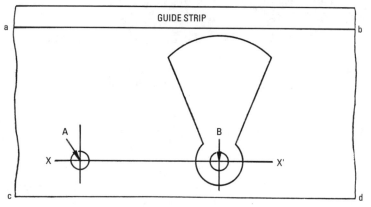

Figure 6-31 Layout of pattern to locate the gage pin.

this distance, measure the distance on a line between corresponding points on the outlines of the two nearest patterns that occupy similar positions in reference to the edge of the stock. Thus, in Figure 6-30, the patterns M and N occupy similar positions in relation to the edge of the stock, as opposed to the pattern P, which is reversed.

The gage pin is then located at a distance *BA* from the point B. Therefore, the distance *BA* (see Figure 6-31) can be laid off on the line XX' equal to the distance BA in Figure 6-30, which fixes the location of the gage pin in reference to the hole in the pattern.

A second layout is necessary to locate the gage pin with respect to some other part of the contour of the pattern (Figure 6-32). If the gage pin is to be located within a space limited by the points r and t, draw the lines aa', bb', and cc' parallel to the edge of the stock and through the points r, s, and t. The points r', s', and t' on the face of the gage pin correspond to the points r, s, and t on the pattern, and they must be located at a distance AB (determined previously in Figure 6-31) from the points r, s, and t, along the parallel lines aa', bb', and cc'.

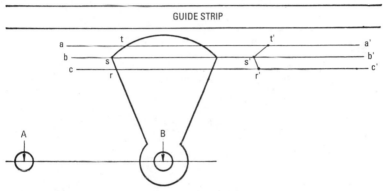

Figure 6-32   A second layout for location of the gage pin with respect to some other point in the outline of the pattern.

The pattern P (see Figure 6-30), which is placed 180° with respect to the patterns M and N, can be punched by turning the sheet of stock 180° and passing it through the press again. The best method of doing this in production work is to use a double-blanking die (that is, a die that can punch the patterns M and P in a single operation).

Laying Out the Design on the Die
The surface of the stock from which the die is to be made should be finished smooth by either filing or grinding. Then, the surface should be treated so that the scribed marks will be visible in transferring the design to the die.

The surface should be free from grease. A solution of one part copper sulfate and 10 parts water can be used to coat the surface. After a few minutes, the solution will evaporate, leaving the surface covered with a thin film of copper, which serves to make finely scribed lines visible because of the difference in color between the steel and the copper.

Making the Die

The simple design (Figure 6-33) can be used to illustrate the operations necessary to make the opening in a blanking die. The large circular ends A and B should be bored out with a drill of the same diameter. Then, several small holes can be drilled to outline the intermediate part of the design. Alternate small holes should be drilled first. Then, the remaining holes should be drilled so that the entire metal section C can be removed from the design without difficulty.

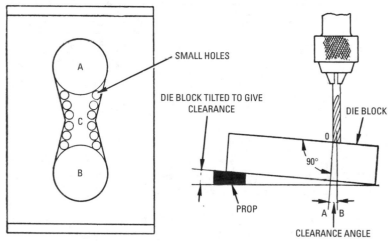

Figure 6-33 Drilling operations for making the opening in a blanking die (left), and method of tilting the die block to obtain clearance (right).

A blanking die must have clearance (see Figure 6-33). Clearance is the taper given the walls of the holes in a die; the large end of the hole is at the bottom of the die. This permits the punched blanks to fall out of the die. In Figure 6-34, a die having no clearance is compared with a die having an exaggerated clearance.

Several methods can be used to obtain clearance. If there were no clearance, the punched blank would remain stuck in the die block (Figure 6-35). The blank is free to fall out of the die block if the die has clearance.

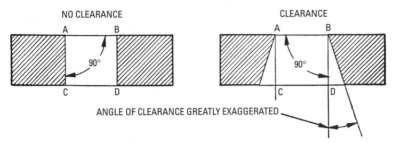

Figure 6-34 Sections of die blocks having no clearance (left) and blocks having clearance (right).

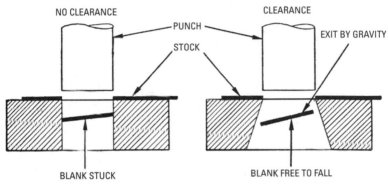

Figure 6-35 Comparison of clearing actions of a die with no clearance (left) and a die with proper clearance (right).

In drilling to the outline of the die design, clearance can be obtained by inserting a prop or thin strip of metal beneath the block on the side farthest removed from the hole that is to be drilled (see Figure 6-33). This tilts the die block so that the drilled hole will be inclined from the vertical at an angle equal to the clearance angle *AOB*.

The holes can be drilled at an angle of 90° to the face of the die block and then reamed from the bottom with a tapered reamer to provide the clearance angle. The opening at the bottom can also be made larger by filing the die opening to the scribed lines.

Usually, a clearance angle of 1° to 2° must be provided. The *diemakers' square* (Figure 6-36) can be used to check for uniform clearance around the die opening. The diemakers' square differs from the real square in that the angle of the blade with the stock of

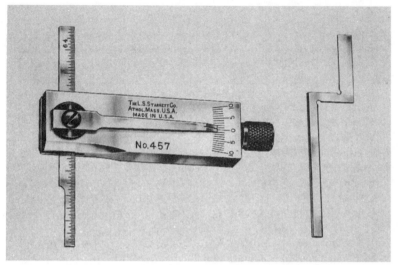

Figure 6-36 A diemakers' square, used for measuring die clearances. The blade is locked in position by a small clamp screw and can be set at any angle up to 8° on either side of 0°. The offset blade is used where it is impossible to sight with the straight blade. *(Courtesy L. S. Starrett Co.)*

the "square" is 92° (90° + 2°) to provide for the clearance angle. The blade is very narrow for accessibility to small openings.

Surplus metal is removed from large dies by machining. It can be removed from the smaller dies with a sharp, cold chisel after drilling. Clearance should not extend entirely upward to the cutting edge (upper face) of the die; it should extend only to within about ⅛ inch of the edge. This is because the die opening would be increased by sharpening, not because of the cutting action of the die.

Hardening and Tempering

The die is ready for hardening and tempering after the preceding operations have been completed. Exceptional skill and judgment are required to harden dies properly.

All holes, such as screw holes for the guide strips and the hole for the gage pin, should be plugged before beginning the hardening operation. Fire clay or asbestos can be used to plug these holes.

It is essential that a uniform heat be maintained during the hardening process. The die should be heated very slowly because the edges and portions having lesser quantities of metal will heat more rapidly than the rest of the die. If the die is heated unevenly, the die will contract unevenly during quenching, which tends to cause the die to crack. The die should be left in the quenching solution until it cools to room temperature. Strong salt brine can be a good quenching solution because it will harden the die satisfactorily at low heat. Salt water or brine will harden a die more satisfactorily at a lower temperature because the brine has a much greater conductivity than fresh water; consequently, it will absorb the heat much faster from the die. Because the hardening of steel depends on the rapidity with which the heat is removed from the steel, it is reasonable to assume that a given degree of hardness can be obtained at a lower temperature when a brine solution is used as a quench.

After the die has been hardened, it should be brightened on its upper surface and tempered. The die can be tempered by placing it on a heated iron plate ($\frac{1}{4}$ to $\frac{3}{8}$ inch in thickness) with the bottom side of the die block downward on the plate. Uneven heating can be avoided by constantly moving the die. The color for drawing depends on the kind of stock to be punched. Usually, the die should be quenched when the face of the die turns a deep straw color. Dies that are used for light-duty work can be harder than those used for heavy-duty work.

Summary

The die consists of the die block and the punch and is used to cut or shape sheet metal or similar materials. The die block is the part of the die that contains a hole that has the same outline as the piece that is to be punched. The punch fits into the hole in the die block, which is the part to which power is applied for performing the cutting operation.

Dies can be classified with respect to the basic operation that they perform (such as cutting dies, shaping dies, and combined cutting and shaping dies). The cutting dies are the simplest form. The plain or blanking die is the type most frequently used. Depending on the shape of the punch, the blank can have any shape.

Various combinations of punches and dies can be used to perform both the cutting and shaping operations. Double-action dies, plain drawing dies, redrawing, gage and follow, and compound dies are just a few used in these operations. Two concentric punches are used in the double-action die. An outer punch is used to perform the operation of cutting and holding the blank. The inner punch is used as a forming or drawing punch. The plain drawing die is intended only for shallow drawing since there is little or no pressure on the blank.

Complicated sheet metal parts are often made by the use of progressive dies. Die clearances are checked with the diemakers' square. Heat-treating a die requires exceptional skill and judgment.

Review Questions

1. What is a die?
2. What is a die block?
3. What is a punch?
4. What is a stripper plate?
5. What is a guide strip?
6. What is a gage pin used for?
7. The shank attaches the punch to the _____ of the press, and the collar of the punching block takes the thrust.
8. How does a self-centering die work?
9. What type of die is used to change the shape of the stock without cutting?
10. The _____ die is a special form of bending die in which part of the stock is bent to form circular beads around the edge of drawn cylindrical parts.
11. The basic principle of operation of the _____ die depends on the action of a springy material placed under pressure.
12. What is a double-action die?
13. Drawing dies usually employ a _____ to shape or draw the work.
14. A _____ die is one in which two or more punches are combined in a single punching head.
15. A _____ die is a modified gang die in which several successive operations are performed progressively.
16. What is the difference between a compound die and a plain or follow die?
17. Identify the following terms:
    a. Burnishing die
    b. Combination die
    c. Trimming die
    d. Reducing die

    e. Performing die

    f. Fluid die

    g. Triple-action die

    h. Progressive die

18. Clean soap and water are recommended as a ____ for drawing brass and copper.

19. The ___ pin is located so that it will hold material in the correct position for punching.

20. The _____ square can be used to check for uniform clearance around the die opening.

Chapter 7

Grinding

Specific grinding machines are made for certain classes and sizes of work. Production grinders are of four main types: cylindrical center, cylindrical centerless, internal, and surface-grinding machines. The tool and cutter, thread, gear, and cam grinders are other types of grinding machines for special purposes.

Note
Refer to *Audel Machine Shop Basics: All New Fifth Edition* for a complete discussion of abrasives, grinding wheels, the sharpening of drills, and the sharpening of lathe and planer tools. This chapter concentrates on the use of grinding wheels on various types of grinders.

Cylindrical Grinders

Most external grinding of cylindrical parts is performed on *cylindrical grinders*. Cylindrical grinding designates a general category of grinding methods with the common characteristic of rotating the work around a fixed axis while grinding surface sections in relation to the axis of rotation. The part being ground is frequently cylindrical (hence the naming of the general category). Straight cylinders, cylinders having more than one diameter, and tapered parts can be ground on these grinders (Figure 7-1). Even a surface of curvilinear profile may be ground as long as the condition of a common axis of rotation is satisfied. Cylindrical grinding is applied to the manufacturing of different sizes of work, from miniature parts to rolling mill rolls weighing several tons. Pieces of work having irregular profiles (such as cams or eccentrics) are ground either on a cam grinder or on a cylindrical grinder with a cam-grinding attachment.

Cylindrical work is usually held between centers. Live centers are used when the work is of such a nature that it would score either center, but two dead centers are used wherever possible. The centers are held in a headstock and tailstock, both mounted on a table that can be moved back and forth in front of the wheel.

Cylindrical grinding machines are usually divided into three categories: plain, universal, and limited-purpose. A *plain cylindrical grinding machine* is the basic type in this general category. It is used for grinding parts with cylindrical or slightly tapered form.

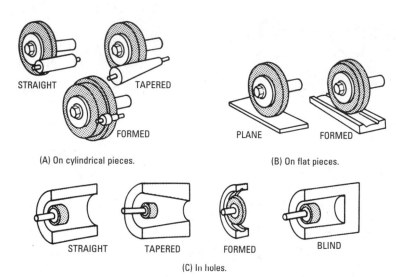

(A) On cylindrical pieces.

(B) On flat pieces.

(C) In holes.

Figure 7-1 Types of grinding commonly performed on external, internal, and flat work. Centerless grinding, either external or internal, is also performed on cylindrical work. *(Courtesy Cincinnati Milacron Co.)*

The *universal cylindrical grinding machine* (Figure 7-2) is a flexible machine. Figure 7-3 shows a woodcut of the original universal grinder invented by J. R. Brown in 1876. The fundamental principles of the machines are the same. The most noticeable difference is the use of electric motors instead of the belt drives. In addition to the basic cylindrical forms, the universal grinder is useful for grinding steep tapers, face grinding, and even internal grinding. Tapers can be ground by swiveling the table (shown in Figure 7-4), by swiveling the headstock, or by the use of special attachments such as a center grinding attachment. The wheel head can be swiveled and the wheel dressed to grind to a shoulder or, in some cases, both the cylinder and a small shoulder (shown in Figure 7-5). Such variety is typical of work in the tool room, where most of these machines are installed.

Limited-purpose cylindrical grinders are basically production machines designed for special work and for high-volume production. Examples include crankshaft grinders, camshaft grinders, roll grinders, and so on.

If the work is long and slender, workrests may be necessary to support the work against the feed or pressure of the grinding

Figure 7-2 Valumaotor universal grinder. *(Courtesy Brown & Sharpe Manufacturing Co.)*

wheels. They should be spaced evenly, from six to ten work diameters apart. The infeed and other basic data are listed in Table 7-1. The length of the surface to be ground is generally limited only by the movement of the machine table. In some larger roll grinders, the movement is accomplished by the traverse of the wheel slide along a stationary machine table.

Cylindrical grinding between centers falls into two categories: traverse and plunge grinding. *Traverse grinding* is used where the work is longer than the maximum width of the grinding wheel that can be mounted on the machine. The worktable on which the headstock and tailstock centers are located reciprocates past the grinding wheel. As the work passes the wheel, a fixed amount of metal (infeed) is removed from the diameter. The grinding wheel is advanced another increment of distance, at the end of the pass, to remove the same amount of stock on the succeeding pass.

Table 7-1   Basic Process Data for Cylindrical Grinding

Traverse Grinding

Work Material	Material Condition	Work Surface Speed, ft/min.	Infeed, in./Pass		Traverse for Each Work Revolution, in Fractions of the Wheel Width	
			Roughing	Finishing	Roughing	Finishing
Plain carbon steel	Annealed	100	0.002	0.0005	½	⅙
	Hardened	70	0.002	0.0003 to 0.0005	¼	⅛
Alloy steel	Annealed	100	0.002	0.0005	½	⅙
	Hardened	70	0.002	0.0002 to 0.0005	¼	⅛
Tool steel	Annealed	60	0.002	0.0005 max.	½	⅙
	Hardened	50	0.002	0.0001 to 0.0005	¼	⅛
Copper alloys	Annealed or cold drawn	100	0.002	0.0005 max.	⅓	⅙
Aluminum alloys	Cold drawn or solution treated	150	0.002	0.0005	⅓	⅙

Plunge Grinding

Work Material	Infeed per Revolution of Work, in.	
	Roughing	Finishing
Steel, soft	0.0005	0.0002
Plain carbon steel, hardened	0.0002	0.000050
Alloy and tool steel, hardened	0.0001	0.000025

Information provided by Machinery's Handbook, The Industrial Press.

Figure 7-3 Original universal grinder. *(Courtesy Brown & Sharpe Manufacturing Co.)*

Figure 7-4 Taper grinding on the universal grinding machine.
(Courtesy Cincinnati Milacron Co.)

Figure 7-5   Grinding to a shoulder. *(Courtesy Brown & Sharpe Manufacturing Co.)*

Figure 7-6 shows a workpiece supported by two such workrests on a universal grinding machine. A simple method is to locate the first rest in the center of the work and use an equal number of rests on either side.

Plunge grinding is accomplished by using a grinding wheel that is at least as wide as the area to be ground. Grinding is accomplished by feeding the grinding wheel into the work, which is mounted between centers and does not reciprocate past the wheel. The infeed is referred to as the "infeed rate" because the infeed of the wheel is continuous rather than intermittent. This infeed rate, as shown in Table 7-1, is measured in ten-thousandths to millionths of an inch per revolution of the work.

Plunge grinding is more widely used in mass production grinding because it is more rapid and lends itself to automatic operation. These grinding wheels are usually ordered to specific thicknesses and are usually trued on the sides (as well as the face) to maintain the precise thickness. Production grinding machines are often built to mount more than one grinding wheel for plunge grinding several diameters on a single shaft simultaneously.

Figure 7-6 Workrests being used on a universal grinding machine.
(Courtesy Brown & Sharpe Manufacturing Co.)

Cylindrical grinders are frequently fitted with grinding gages. These unique instruments consist of a dial indicator that reads the diameter of the work while the surface is being ground. The dial indicator can read in ten-thousandths of an inch, which allows the operator to grind right to size quickly and accurately. The gages can also be used with automatic machine controls. Figure 7-7 shows a gage on the work being ground while Figure 7-8 shows the gage retracted for replacing the workpiece.

Centerless Grinders

Centerless grinding is the method of grinding cylindrical surfaces without rotating the work between fixed centers. This method of grinding is accomplished by supporting the work between three fundamental machine components—the grinding wheel, the regulating or feed wheel, and the work blade. The grinding and regulating wheels rotate in the same direction, establishing a "climb cut" operation.

Centerless grinding has developed into one of the most important operations in industry. The centerless grinding machine (Figure 7-9) of the Cincinnati Milacron Co. can obtain roundness

Figure 7-7   Grinding gage measuring work during grinding.
(Courtesy Federal Products Corp.)

within 0.000050 inch, finishes below 0.000010 inch, and size tolerances within 0.0001 inch at high-production rates.

The centerless grinder can be used on a wide range of materials such as cork, glass, porcelain, wood, rubber, plastics, and the new high-alloy steels, as well as the more common types of ferrous and nonferrous materials. As its name implies, the centerless grinder requires no center points. There is no need to locate and drill center holes in the workpiece.

Basic Principles
The principal parts of a centerless grinder are the grinding wheel, regulating wheel, and workrest. Suitable guides are used in the

Figure 7-8 Grinding gage retracted for replacing workpiece.
(Courtesy Federal Products Corp.)

Figure 7-9 Centerless grinding machine. (Courtesy Cincinnati Milacron Co.)

"throughfeed" workrest to lead the workpiece to the wheels and to receive it from the wheels, as well as proper means for supporting the workpiece during the grinding cut. All these elements can be arranged and combined in various ways, but the fundamental principle involved is always the same.

Action of Grinding and Regulating Wheels

The action of the grinding wheel forces the workpiece against the workrest by means of *cutting pressure* and also against the regulating wheel by *cutting contact pressure*. The cutting pressure, aided by the force of gravity of the workpiece, keeps the workpiece in contact with the regulating wheel. The regulating wheel is usually made of material that is similar to the grinding wheel, and it provides a continuously advancing frictional surface that ensures constant and uniform rotation of the workpiece at the same velocity as the periphery of the regulating wheel.

It is rather simple to understand how a cylindrical surface can be ground on a center-type grinder. The diameter of the cylinder is governed by the distance between the line of centers and the face of the grinding wheel. However, there are no centers on the centerless grinder, and no apparent method of controlling the roundness of the workpiece exists. The ground diameter of the workpiece is determined by the distance between the two active surfaces of the wheels. However, a constant diameter does not necessarily indicate a perfect cylinder.

Figure 7-10 shows one of the simplest setups for centerless grinding. The center of the workpiece is in line with the centers of the grinding wheel and the regulating wheel. A work blade with a flat top is used to support the workpiece. The surfaces of the grinding wheel and the regulating wheel, together with the flat work blade, form three sides of a square.

Figure 7-11 shows the effect of an out-of-round piece of work placed "on center" on a work support blade with a flat top. A high spot on the periphery of the workpiece produces a diametrically opposed concave spot when the high spot comes in contact with either the grinding wheel or the regulating wheel. A grinding setup of this type produces a piece of work that has a constant diameter, but the workpiece is not cylindrical. Figure 7-12 shows an exaggerated shape generated by this type of setup.

To correct the effect of an out-of-round workpiece, the center of the workpiece can be elevated above the centerline of the wheels by raising the work support blade (Figure 7-13). Then, a low spot on the workpiece coming in contact with the regulating wheel causes a

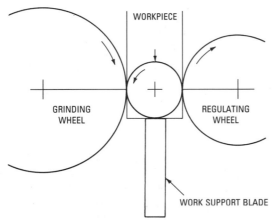

Figure 7-10 Centerless grinding operation. Note that the center point of the workpiece is placed "on center," and the work is supported by a blade with a flat top.

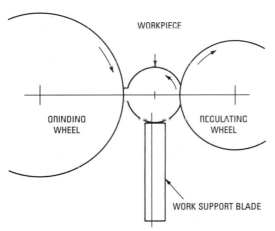

Figure 7-11 The effect of centerless-grinding an out-of-round workpiece placed "on center" on a work support blade with a flat top.

high spot to be generated at the point of contact with the grinding wheel, but the spots are not diametrically opposite. As the rotating piece is being ground, the low and high spots will not be opposite, and a gradual rounding effect is obtained. Maximum corrective action can be obtained by using a work support blade that is inclined at the top, as shown in Figure 7-13.

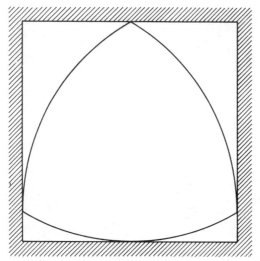

Figure 7-12 An exaggerated shape resulting from centerless-grinding an out-of-round workpiece placed "on center" on a work support blade with a flat top.

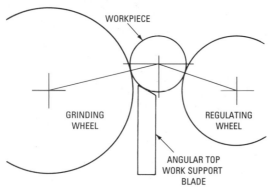

Figure 7-13 Elevating the workpiece above the centerline of the grinding and regulating wheels to correct the rounding action produced in grinding an out-of-round workpiece.

Figure 7-14 shows a diagrammatic sketch of the setup for corrective action in grinding an out-of-round workpiece. The two lines *AA* and *BB* are tangent at the points of contact of the workpiece with the grinding wheel and the regulating wheel, and another line *CC* indicates the plane of the inclined top of the work support

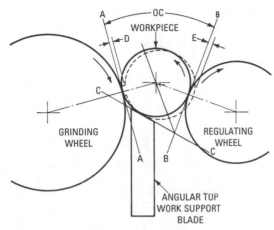

Figure 7-14 Diagrammatic sketch of setup for corrective action in grinding an out-of-round workpiece.

blade. If a low spot on the workpiece contacts either the work support blade or the regulating wheel (indicated by dotted lines), the approximate center point of the workpiece is lowered. Because the center points of both wheels are fixed, the smallest possible diameter is generated only when the center point of the workpiece is located on their centerline.

Any increase in height of the center point of the work above the centerline of the wheels causes the generation of a workpiece with a larger diameter, and lowering the center point of the work toward the centerline of the wheels reduces the diameter of the ground piece. Thus, the grinding wheel, instead of leaving a high spot on the periphery of the work corresponding to the depth of the concave spot at the point of contact with the regulating wheel, generates a proportionally smaller high spot at its point of contact with the workpiece. Theoretically, the low and high spots "dampen" themselves out, approaching the cylindrical shape in infinity, but the cylindrical form can be obtained in a short time in actual practice.

The position of the work in relation to the two wheels should be such that the center of the work does not fall on the centerline of the two wheels. The work is usually placed above center for fast rounding action. In some instances (especially for long, slender work) it is placed slightly below center to avoid a whipping action of the work.

Centerless grinders are used in shops where relatively large quantities of the same workpieces are ground. Some odd-shaped pieces (which tend to tilt when placed in the machine or to have projections too close to the ground surface) do not lend themselves to centerless grinding. There are other pieces (such as long, slender rods) that can be handled only on centerless machines.

Centerless Grinding Methods

Centerless grinding consists chiefly of four types: throughfeed, infeed, endfeed, and combined infeed and throughfeed.

Throughfeed

With *throughfeed grinding*, there is a fixed relation between the grinding wheel, regulating wheel, and the work support blade (Figure 7-15). The regulating wheel can be adjusted so that the distance between the active surfaces of the wheel, together with the height of the work support blade, determines the diameter of the ground piece. The regulating wheel can be readjusted slightly to compensate for grinding wheel wear and restore the original relation between the wheels and support blade, which determines the size of the ground workpiece. The workrest, which supports the blade, is provided with adjustable guides both at the front and at the rear of the grinding wheel and regulating wheel. These guides direct the workpiece in a straight line to and from the wheels. Each of the guides can be adjusted by means of screws to the correct relation between the wheel surfaces and the diameter of the work.

In throughfeed grinding operations, the work is passed between the grinding and regulating wheels, and the grinding action takes

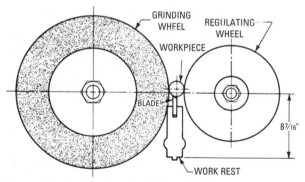

Figure 7-15   The height of the wheel center points above the bottom of the workrest. The height of the center point of the workpiece relative to the centerline of the wheels can be calculated by subtracting 8⁷⁄₁₆ inches from the distance between the center point of the workpiece and the finished surface of the lower side.

place as the workpiece passes from one side of the wheels to the other side. Only *straight cylindrical surfaces without interfering shoulders* can be ground by this method because all points on the workpiece pass all contact points between the wheels (Figure 7-16).

The regulating wheel imparts the axial movement of the workpiece past the grinding wheel. The regulating wheel can be swiveled

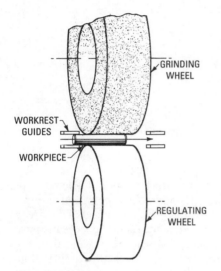

Figure 7-16   Top view of a throughfeed centerless operation.

GRINDING WHEEL

WORKREST GUIDES

WORKPIECE

REGULATING WHEEL

about a horizontal axis from 2° below to about 8° above a line relative to the axis of the grinding wheel spindle. The speed and diameter of the regulating wheel also influence the feeding rate of the work.

Infeed

The *infeed centerless grinding* operation corresponds to plunge-cut or form grinding on the center-type grinder. This method can be used to grind a workpiece *that has a shoulder, head, or other portion that is larger than the ground diameter* (Figure 7-17). The infeed method can be used for simultaneous grinding of several diameters of a workpiece, as well as for finishing workpieces that have an irregular profile. The length of the sections that can be ground in a single operation is not limited by the width of the grinding wheel. The combined infeed-throughfeed method can be used to grind work lengths that are longer than the width of the wheels.

There is usually (but not always) a fixed relation between the regulating wheel and the work support blade. The units that

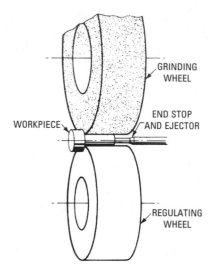

Figure 7-17 Top view of an infeed centerless grinding operation.

GRINDING WHEEL

END STOP AND EJECTOR

WORKPIECE

REGULATING WHEEL

incorporate the regulating wheel and the work support blade are supported by two slides, which are clamped together (upper side clamps) and carry the workpiece to and from the grinding wheel. The infeed lever is turned 90° to perform this movement. As the lever is brought downward, the work and the regulating wheel are advanced to the grinding wheel, securing the desired size as the lever has completed its full swing. The gap between the wheels is increased as the movement of the infeed lever is reversed, and either a manually or automatically operated ejector kicks the work outward from between the wheels. Then, another piece can be placed in position by the operator.

Form grinding is a good example of the type of infeed or plunge grinding on centerless machines. Figure 7-18 illustrates this type of work. Out-of-balance work can sometimes be accomplished by the use of a hold-down shoe, as shown in Figure 7-19.

Endfeed
The *endfeed* method of centerless grinding is used only on tapered workpieces. Either the grinding wheel or the regulating wheel, or both, must be dressed to the correct taper (Figure 7-20). The grinding wheel, regulating wheel, and work support blade are set in fixed relation to each other. The workpiece is fed inward from the front, either manually or mechanically, to a fixed end stop.

Combined Infeed and Throughfeed
The infeed and throughfeed can often be combined (Figure 7-21) for three chief types of applications as follows:

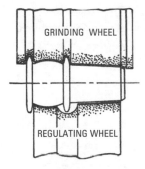

Figure 7-18 Plunge grinding multiple-diameter work.

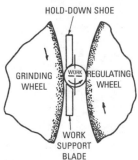

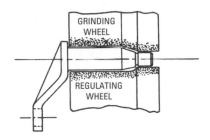

Figure 7-19 Plunge grinding out-of-balance work using a hold-down shoe. *(Courtesy Norton Co.)*

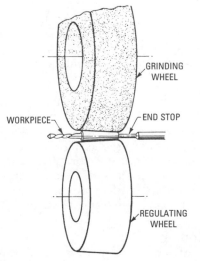

Figure 7-20 Top view of an endfeed centerless grinding operation.

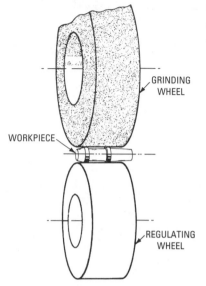

Figure 7-21 Top view of combined infeed and throughfeed operation.

GRINDING WHEEL

WORKPIECE

REGULATING WHEEL

- For grinding parts that can be ground more conveniently in a single pass but are too large for the conventional throughfeed method.
- For grinding the smaller diameter of parts that have two diameters, where the portion that is to be ground exceeds the width of the grinding wheel.
- For warped parts in which the bow or warpage does not exceed the total amount of stock to be removed, and the length of the part is less than the width of the grinding wheel.

Abrasive-Belt Centerless Grinding

Centerless grinding machines for cylindrical work have been designed using abrasive-belt machines. Contact with the work is made at the point where the abrasive belt runs around one of the pulleys. The work is backed up by a regulating wheel. This wheel controls the rotation and forward movement of the work just as in other centerless grinding machines. The finish and accuracy are comparable to the abrasive-wheel centerless grinders for many types of straight work, and the belt machines are much less expensive.

Advantages of Centerless Grinding

Some of the chief advantages offered by centerless grinding machines are as follows:

- Compared with center-type grinding, the loading time is small and enables the grinding process to be practically continuous.

- The workpiece is rigidly supported directly beneath the grinding cut and for the full length of the cut. There is no deflection during the cut, and a heavier cut can be taken if desired.

- Long, brittle workpieces and easily distorted workpieces can be ground because there is no axial thrust on the workpiece, as compared with center-type grinding.

- The error of centering the workpiece is eliminated, because a true floating condition is present during the grinding process.

- Possible error in setting up the job is reduced because stock removal is measured on the diameter rather than on the radius; likewise, error due to wheel wear is reduced.

- Maintenance is reduced because there are fewer wearing surfaces on the machine.

- Larger quantities of small workpieces can be fed automatically by means of a magazine, a gravity chute, or hopper feeding attachments.

- Extremely accurate control of size in production can be attained.

- Very little skill is required by the machine operator.

- Large grinding wheels can be used so that wheel wear is minimized.

Internal Grinding

Internal grinding machines are used for finishing holes to accurate diameters in such parts as bushings, gears, bearing races, cutters, and gages. Internal grinding may be done either on universal grinding machines or on machines especially designed for that purpose. Figure 7-22 shows a setup for internal grinding on a universal grinding machine. The work is held in a chuck on the headstocks while the grinding wheel is mounted on an internal grinding spindle driven by a flat belt.

Rotating-Work Machine

In a machine of this type, the work is held in a chuck on the spindle of the working head. The wheel spindle moves in and out of the hole during the grinding. Either straight holes or tapered holes may be ground because the headstock is made to swivel to various angles on either side of the centerline.

Figure 7-22   Internal grinding on a universal grinder. *(Courtesy Brown & Sharpe Manufacturing Co.)*

Internal Centerless Grinding Machine

Special machines have been developed for centerless internal grinding. The process is usually automatic, and the ground hole will be concentric with the outside diameter of the work. The machines can grind straight cylindrical or tapered holes. The holes may be blind, through, interrupted, or with a shoulder.

Figure 7-23 illustrates the principle. The process involves three rollers to hold the work and cause it to rotate on its outer surface.

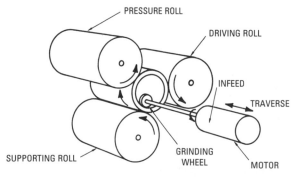

Figure 7-23   Principle of centerless internal grinding.

One roller is the *regulating roller*, which drives the work and controls its speed and direction. The second roller is mounted below the work to support it and controls the distance from the work center. The third roller is a *pressure roller* and holds the work in contact with the other two. This roller moves in and out to allow for loading and unloading the machine. The grinding wheel is usually in a fixed position, and the work moves back and forth longitudinally. When two or more grinding operations must be performed on the same part (such as roughing and finishing), the work can be rechucked in the same location as often as required. The centerless grinding method simplifies the setup and permits automatic loading, so it is necessary only to keep the loading magazine full and to take away the ground parts. Figure 7-24 shows two different procedures for grinding the inside of thin wall sections.

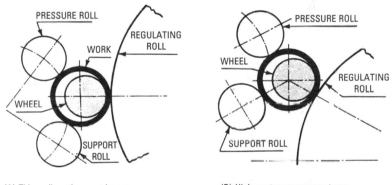

(A) Thin wall sections can be centerless ground internal without distortion by an on-center arrangement of the rolls.

(B) High-center arrangement permits maintenance of high grinding accuracy, even with considerable variation in outside diameter.

Figure 7-24　Internal centerless grinding of thin wall sections.
(Courtesy Cincinnati Milacron Co.)

Cylinder Grinding Machine—Stationary Work

Sometimes called the *planetary type* of internal grinder, this machine was developed originally for the automobile industry for grinding cylinder block bores. The work is mounted on the table of the machine and travels to and from the wheel, which not only rotates on its axis but also travels in a circular path of any desired diameter. This type of machine is now used for bore grinding in heavy or irregularly shaped work that cannot be readily rotated.

Surface Grinders

The operation of producing and finishing flat surfaces with a grinding machine is called *surface grinding* (Figure 7-25). The term *surface grinding* is generally accepted as meaning the grinding of plane surfaces only. Surface grinders vary in size from small toolroom

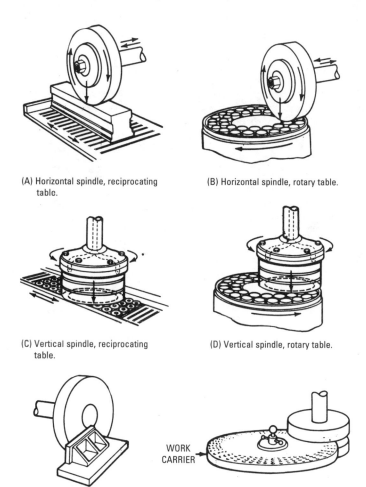

(A) Horizontal spindle, reciprocating table.

(B) Horizontal spindle, rotary table.

(C) Vertical spindle, reciprocating table.

(D) Vertical spindle, rotary table.

(E) Horizontal spindle, single disc.

WORK CARRIER

(F) Vertical spindle, double disc.

Figure 7-25 Principal types of surface grinding include flat and rotary tables and horizontal and vertical wheels. Disc grinders may be single- or double-wheel and either vertical or horizontal. *(Courtesy Cincinnati Milacron Co.)*

machines using 7-inch × ½-inch wheels to very large grinding machines with 48-inch to 60-inch segmental chucks (Figure 7-26).

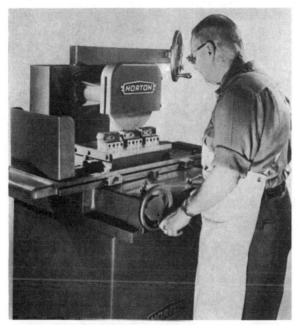

Figure 7-26   Hand-operated surface grinder. *(Courtesy Norton Co.)*

Surface grinders can be classified in two different ways—by the type of table movement or by the position of the spindle. The table movement is either *reciprocating* or *rotary*. The position of the spindle is either vertical or horizontal, and either type can have a reciprocating or rotary table.

Surface grinding requires that the spindle runs true so that there is constant pressure between the wheel and the work. If the spindle bearings are loose, vibration of the spindle and wheel can happen, which does not result in precision grinding.

The infeed or depth of cut is controlled by the amount of stock to be removed, the desired finish, the operating temperatures, and the power capacity of the machine. A coolant can be used to reduce the operating temperature. An ample flow of coolant is more important than its pressure of application. Some vertical spindle machines have the coolant fed through the hollow spindle to the inside of the grinding wheel. Wheel speeds are usually lower on vertical spindle machines because the area of contact between the work and the wheel is greater.

Planer-Type Surface Grinders

The worktable reciprocates in the *planer type of surface grinder* (Figure 7-27). There are three types of planer-type surface grinders. One uses the outside diameter (or periphery) of the grinding wheel mounted on a horizontal spindle; another uses a cup, cylinder, or segmental-type wheel mounted on a horizontal spindle; and a third uses a cup, cylinder, or segmental-type wheel mounted on a vertical spindle at right angles to the work.

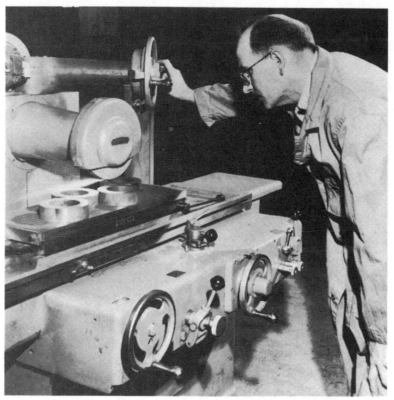

Figure 7-27 Horizontal-spindle, reciprocating-table surface grinder.
(Courtesy Norton Co.)

Rotary-Type Surface Grinders

There are two classes of *rotary surface grinders*. One uses the periphery of the grinding wheel mounted on a horizontal spindle

(Figure 7-28), and the other uses a cup, cylinder, or segmental wheel mounted on a vertical spindle.

Figure 7-28 Rotary surface-grinding machine. Note that a taper-shank gear cutter is in position for grinding. To accommodate the shank, an adapter and support plate are mounted on the magnetic chuck, centering on a locator in the pilot hole. *(Courtesy Heald Machine Co.)*

Cutter and Tool Grinding

A cutter deteriorates rapidly if it is used after it becomes dull. Its life will be prolonged if it is sharpened at the proper time because very little stock needs to be removed or ground off each time. The life and efficiency of a twist drill are greatly dependent on proper sharpening. Sharpening of lathe and planer tools likewise should receive careful attention and should be done by trained operators on suitable equipment with properly specified grinding wheels. It is especially important that multiple-point cutting tools, such as taps, milling cutters, hobs, reamers, and so on, be sharpened properly.

Grinding Cemented Carbide Tools

The technique of grinding these tools has been developed to the point where they are considered no more costly to grind than high-speed steel tools. The expanded use of cemented carbides since World War II has emphasized the importance of correctly sharpening tools and cutters tipped with the hardest of cutting alloys.

Cutter Sharpening Machines

These machines are usually divided into two categories: plain cutter grinders and universal tool and cutter grinding machines. The *plain cutter grinder* is especially suitable for sharpening milling machine cutters and similar tools.

The *universal tool and cutter grinding machines* have sufficient range to meet practically all toolroom requirements. Basically, they are similar to small universal cylindrical grinders with four important differences:

- No power feeds are provided; all table motions are manual.
- The headstock can be swiveled about a horizontal axis as well as a vertical axis.
- There is no motor on the headstock.
- The wheelhead can be raised and lowered, as well as swiveled 360° about a vertical axis.

The high degree of flexibility built into this machine makes it possible to grind almost any type of tool.

The grinding of cutting tools is primarily a problem of holding and moving the cutting tool in proper relationship to the grinding wheel. The setup of the grinder is a complicated job and requires a skilled worker. However, the actual grinding is rather easy after the machine is set up for a particular job. Figure 7-29 shows the setup for grinding a cutter mounted on a mandrel. Sharpening such a cutter requires that the rake and clearance angles be correct. These are checked by use of a cutter clearance gage. Figure 7-30 shows this gage being used to check the clearance angle of a milling cutter. Figure 7-31 shows the same gage being used to check the clearance angle of a broach.

Carbide tools can be ground readily either with silicon carbide abrasive wheels (Figure 7-32) or with diamond wheels, which may be resinoid, metal, or vitrified bonded. Aluminum oxide wheels are not suitable for grinding cemented carbide and should never be used on carbide tools, except for undercutting the steel shank.

Figure 7-29 Grinding a cutter on a universal tool and cutter grinder.
(Courtesy Brown & Sharpe Manufacturing Co.)

Diamond grinding wheels are the accepted type of abrasive wheel for offhand finish grinding of carbide single-point tools and for all fixed-feed or precision grinding operations on cemented carbides, including grinding of chip breakers and multitooth cutters. They have the advantages of fast and cool cutting action and extremely low rate of wear compared with silicon carbide wheels. These advantages more than offset their higher price and make them more economical for these operations.

Diamond wheels, exclusively, are recommended for sharpening carbide-tipped face mills, reamers, and other multitooth cutters (Figure 7-33). The wheels may be either vitrified or resinoid bonded.

Barrel Finishing (Abrasive Tumbling)

Barrel finishing is a precisely controlled method of processing quantities of parts, both metallic and nonmetallic, to remove sharp edges, burrs, machining or grinding lines, and heat-treat scale and

Figure 7-30 Checking milling cutter clearance angle. *(Courtesy L. S. Starrett Co.)*

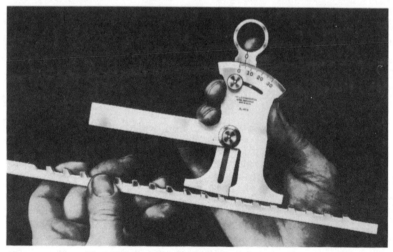

Figure 7-31 Checking broach clearance angle. *(Courtesy L. S. Starrett Co.)*

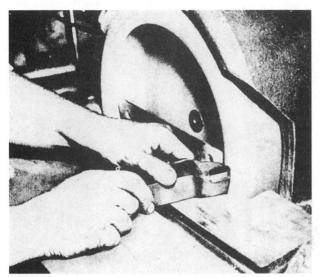

Figure 7-32 Sharpening a single-point carbide planer tool on a silicon carbide grinding wheel. *(Courtesy Norton Co.)*

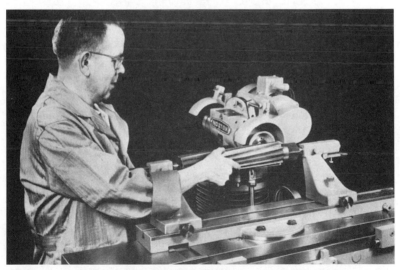

Figure 7-33 Sharpening a reamer on a cutter and tool grinder. *(Courtesy Norton Co.)*

to improve surface finish (Figure 7-34). The process may be used preliminary to buffing or plating operations. It may provide the desired finish for end use of the working parts for such machines as typewriters and business machines. The process is a modern quantity production method that results in savings in man-hours and cost per part manufactured.

Figure 7-34 Before and after barrel finishing. *(Courtesy Norton Co.)*

Typical parts being finished commercially by the barrel-finishing method are stampings, die castings, small sand castings, machined parts, forgings, and sintered metals. Materials that can be barrel finished include steel, brass, cast iron, copper, bronze, aluminum, magnesium, titanium, silver, plastics, rubber, agate, and glass (Figure 7-35).

Almost all barrel finishing in abrasive media is done wet. Following are the essential components of the barrel-finishing process:

- A rotating barrel or vibration-type unit
- Abrasive media (a nonabrasive may be used for some purposes)
- A chemical compound
- Water

The basic action of barrel finishing is a sliding movement of the upper layer of the workload as the barrel rotates (Figure 7-36). This action consists of a rotary movement of the mass of abrasive and parts to the point where the pull of gravity overcomes the tendency of the mass to stay together. The upper layer of the load then slides

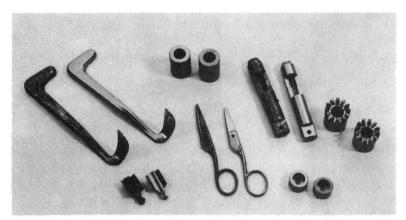

Figure 7-35 Samples of typical parts before and after barrel finishing. *(Courtesy Norton Co.)*

(neither falls nor is thrown) toward the lower portion of the barrel. Nearly all (85 to 95 percent) of the abrading, deburring, or polishing occurs during the sliding action. Little or no cutting action takes place while the parts are moving up through the mass from the bottom of the barrel to the point where the sliding starts. Grinding, deburring, descaling, polishing, or honing to a mirror finish can be secured by tumbling. The size and shape of the barrel, the speed of rotation, the type of abrasive and lubricant, the proportion of parts and abrasive, the length of time, and the material and shape of the part being tumbled must be considered to secure the finish desired.

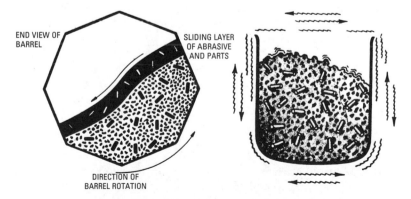

Figure 7-36 Barrel-finishing action within a rotating barrel (left) and a vibratory barrel (right). *(Courtesy Norton Co.)*

Summary

Four main types of production grinders are cylindrical center, cylindrical centerless, internal, and surface-grinding machines.

Cylindrical grinding is usually done between centers. The surface must be concentric around the axis but does not have to be parallel to the axis. Cylindrical grinding can be divided into two types—traverse and plunge grinding. The universal cylindrical grinding machine is quite flexible. Attachments can be used to provide additional operations. Gages can be used to actually measure the work while the part is being ground.

The basic parts of a centerless grinder are the grinding wheel, regulating wheel, and workrest. Various guides are used in the throughfeed workrest to lead the workpiece to the wheels and to receive it from the wheels, as well as a proper means of supporting the workpiece during the grinding cut.

The grinding wheel forces the workpiece against the workrest by means of the cutting pressure and also against the regulating wheel by the cutting contact pressure. The regulating wheel is usually made of material that is similar to the grinding wheel and provides a continuously advancing frictional surface that ensures constant and uniform rotation of the workpiece at the same velocity as the periphery of the regulating wheel.

Centerless grinding consists chiefly of four types: throughfeed, infeed, endfeed, and combined infeed and throughfeed. In throughfeed grinding operations, the work is passed between the grinding and regulating wheels, and the grinding action takes place as the workpiece passes from one side of the wheels to the other side.

The abrasive-belt centerless grinder is an important development for many types of straight work. The finish and accuracy are as good as in the other types of machines, and the belt machines are much less expensive.

Internal grinding is for finishing holes to accurate diameters. The operation can be done on universal grinding machines or on special machines. The centerless internal grinding machine has been a big cost-saver in production. It is fully automatic and can produce small parts with the hole concentric with the outside diameter. And, best of all, it can produce the parts much cheaper than other methods.

Surface grinding involves producing and finishing flat surfaces with a grinding machine. The spindle of a surface grinder can be in a horizontal or vertical position. The table movement can be either reciprocating or rotary.

Tool and cutter grinders are special grinders used to sharpen tools and cutters. The setup is complicated and requires a skilled toolmaker. The actual grinding is not difficult after the setup is made. The rake and clearance angles of tools and cutters need to be measured accurately.

Tumbling is a precisely controlled method of finishing quantities of parts. It is a vital part of the production process.

Review Questions

1. Most _____ grinding of cylindrical parts is performed on cylindrical grinders.
2. Cylindrical work is held usually between _____.
3. List three types of grinding commonly performed on external and internal flat work.
4. When was the original universal grinder invented by J.R. Brown?
5. What are traverse and plunge grinding?
6. Centerless grinding is the method of grinding _____.
7. List the three principal parts of a centerless grinder.
8. What are the four types of centerless grinding methods?
9. Describe infeed centerless grinding.
10. Form grinding is a good example of the type of infeed or _____ grinding on centerless machines.
11. What types of workpieces make use of the endfeed method of grinding?
12. Describe abrasive-belt centerless grinding.
13. List five advantages of centerless grinding.
14. Describe how internal centerless grinding machines do their job.
15. How are surface grinders classified?
16. What is a planer-type surface grinder?
17. Describe the rotary-type surface grinder.
18. What are four differences between the small universal cylindrical grinder and the universal tool and cutter grinder?
19. Carbide tools can be ground readily either with silicon carbide abrasive wheels or with _____ wheels.
20. Describe the process of barrel finishing.

Chapter 8

Laps and Lapping

In making tools, dies, jigs, fixtures, and gages, it is often necessary to provide still finer surface finishes and accuracy for certain surfaces after the grinding operation has been completed. On modern tools, it is often necessary to use a lapping operation to obtain the finish necessary.

Lapping is an abrading process done with a lap charged with an abrasive compound to produce true surfaces, improve dimensional accuracy, correct minor surface imperfections, and provide a close fit between two surfaces. Gage blocks are lapped to plus or minus 0.000002 inch per inch of length, and they are also parallel within this dimension. Very little heat and pressure (which induce strains in the finished part) are involved in lapping. Lapping is more accurate than grinding and some other finishing operations.

Lapping is used on various types of surfaces—flat, cylindrical, spherical, or specially formed. Lapping is accomplished by bringing the workpiece in contact with a lap, usually made of a soft metal. Fine, loose abrasive mixed with a vehicle such as oil, grease, or water is used to charge the lap. Motion between the lap and workpiece is necessary to provide the abrading.

Laps

A *lap* is a tool used for a superfine finish grinding or abrading operation. There are numerous types of laps designed to meet the requirements of various grinding operations.

Classification

Laps can be classified with respect to the degree of contact with the workpiece as *line* and *full* laps, and with respect to their shape as *cylindrical* (either ring or split), *flat*, and *adjustable* (which can be three-step, contracting, or expanding in type). Laps can also be classified with respect to their applications as either *internal* or *external*, and with respect to the material from which they are made (such as lead, copper, brass, cast iron, and steel).

Materials

Various materials are used for laps, depending on the service for which they are intended. A lap made of *lead* can be easily charged with abrasive, holds the abrasive firmly, and does not scratch the workpiece. These laps are relatively easy to fit to the workpiece and hold their shape well for light cuts. They are also suitable for heavy-duty lapping operations.

Adjustment for wear can be provided on laps that are made of lead. The lap is molded to a tapered arbor, which provides for a slight adjustment (Figure 8-1). The arbor should have a keyway to prevent the lap from turning on the arbor. To adjust for wear, the arbor can be driven farther into the lap, thus expanding the lead. Figure 8-2 shows an internal adjustable split lap.

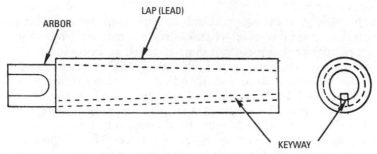

Figure 8-1 An expanding internal lap made of lead (left). The lap is molded on the tapered arbor, which has a keyway (right) to prevent the lap from turning on the arbor.

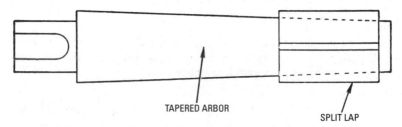

Figure 8-2 An illustration of the internal adjustable split lap.

Of course, the lead will expand only until it fractures. Therefore, the amount of adjustment for wear is quite limited. This expansion of the lead tends toward distortion, and the lap must be either trued by turning or replaced after two or three adjustments. Laps made of lead have a disadvantage in that they tend to lose their form. However, they are inexpensive, easily molded, and quickly charged.

A lap made of close-grained *cast iron* is usually considered the best lap for fine, accurate work. However, some machinists prefer a lap made of *copper* to either cast iron or lead. Laps that are made of copper can be charged more easily, and they cut more rapidly than cast-iron laps, but they do not produce as fine a finish.

Laps are also used to finish holes. Laps made of copper or brass are sometimes used for holes ¼ to ½ inch in size, and laps made of cast iron are used for the larger holes. Laps that are used to finish holes should be longer in length than the length of the hole.

External laps are usually made in the form of a *ring*. The ring forms the holder, and the inner shell is the lap proper. The inner shell can be made of lead, copper, and so on, whichever material is best suited for the purpose for which the lap is intended. The external lap is usually split, and adjustment is made by means of the screws in the holder (Figure 8-3). The lap proper consists of a split ring inside a collar that is provided with a handle. A pointed screw engages the split ring to prevent its turning inside the collar. The remaining two screws are adjusted screws. The external lap should be at least as long as the diameter of the work.

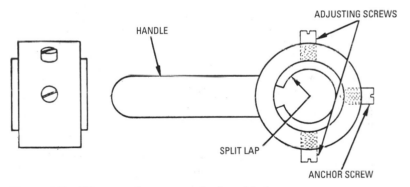

Figure 8-3   Diagram of an external adjustable lap.

Figure 8-4 shows an adjustable external *step* lap for lapping plug gages. This type of lap is usually made of cast iron, and all holes are the same size. In the lapping operation, the first hole A is adjusted to size by means of the adjusting screw. Then, the workpiece is passed from one hole to another during the lapping operation. There is less wear on the lap because of the several holes. The holes retain the same proportions, and they save time in frequent measuring checks when rough-lapping the plug gage to within 0.001 inch of finished size.

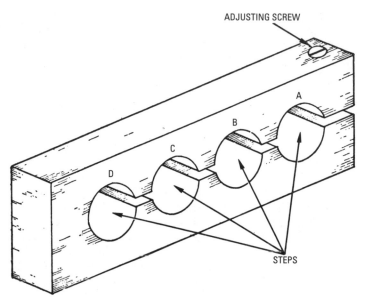

Figure 8-4 An adjustable external step lap, used to lap plug gages.

Lapping Powders

The various "flours" used for lapping usually average about 0.002 inch in diameter. Lapping powders are made in a variety of sizes that indicate the grain or fineness of the abrasive particle used.

The ideal abrasive for lapping is one that will break down (that is, it will become more minutely divided during the lapping operation). For fine work, the abrasives are mixed with oil and allowed to settle. The lighter particles remain in suspension and are poured off. The particles are graded as to the length of time that they float. The most-used lubricants for lapping operations are machine oil, hard oil, kerosene, gasoline, turpentine, and alcohol.

In recent years, diamond lapping powder has come into considerable use because of the necessity for grinding and lapping carbide tools to a fine finish. Diamond lapping powder is made by crushing small diamond chips in a steel mortar and then graining the dust for fineness. A coarse grade of powder is used for roughing tools; a fine grade is used for finishing tools.

Lapping Operations

Lapping is a rubbing process for removing minute amounts of metal from surfaces that must be precision-finished with respect to dimension and smoothness. A piece of work can be brought to a given dimension with greater accuracy and a finer finish than can be obtained by any other means.

Note
Laps were first used by lapidaries to finish the surfaces of mineral specimens, but laps have been in common use for some time in the machine shop to finish fine surfaces.

The more common lapping applications are those operations used to finish micrometer ends, plug and ring gages, holes in jig bushings, and for the finest die work. Lapping operations can be performed either by hand or by machine.

Hand Lapping
An important requirement of hand-lapping operations is that the motion should be in an ever-changing path to obtain uniform abrasion and to eliminate parallel grain marks on both the work and the lap. A matte (dull finish) surface can be obtained by lapping with an ever-changing path.

When lapping a flat surface, a No. 100 or No. 120 emery (or other abrasive of similar grade) and lard oil can be used for charging the roughing lap. After charging, the lap should be washed off and the surface kept moist with kerosene. Gasoline permits faster cutting speed, but the lap will not produce work that is true because of rapid and uneven evaporation of the gasoline.

Flat laps are usually made of cast iron, and they can be scored or grooved so that they will cut better for roughing (Figure 8-5). A plain lap with an unscored surface and charged with a fine abrasive can be used for a fine finish (Figure 8-6).

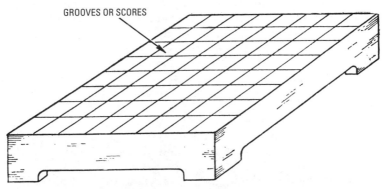

GROOVES OR SCORES

Figure 8-5   A grooved flat lap made of cast iron.

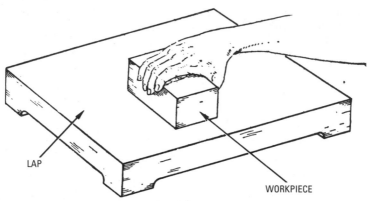

Figure 8-6 A method of flat lapping.

Hand lapping is a slow and tedious operation. The operation is usually performed on flat surfaces by rubbing the parts to be lapped over the accurately finished flat surface of a master lap.

Machine Lapping

Three types of lapping medium are used on lapping machines in industry:

- Metal laps and loose abrasive mixed with a lubricant
- Bonded abrasives for commercial production work
- Abrasive paper or cloth

Vertical lapping machines with cast-iron laps and loose abrasives are used for both cylindrical and flat workpieces. The *universal* lapping machine can also be used for both types of work.

The vertical lapping machines are also used with bonded abrasive laps for both cylindrical and flat work. The horizontal lapping machines use either abrasive paper or cloth.

The workholders differ widely for the vertical lapping machines. The simplest form of workholder is a circular disc that is perforated with holes near its periphery.

These holes are made so that the work can be laid in them as the workholder rests in position on the machine. Sufficient clearance is provided between the workpieces and the workholder for the workholder to move the work about between the laps.

When the workholder is arranged for cylindrical workpieces that do not have a hole through them, the openings are made to the approximate shape of the workpiece. The axis of the opening is not radial, but it is tangent to a circle and concentric with the center point of the workholder. A leg or spider-type workholder is used for cylindrical workpieces (such as piston pins) that have a hole through them.

The centerless lapping machine (Figure 8-7) can be used to improve the accuracy and finish of commercially ground parts. Finishes can be obtained within 1 to 2 microinches; diameter tolerances within 0.00005 inch and 0.000025 inch are possible on the centerless lapping machine. Centerless lapping is particularly suitable for parts ground by the throughfeed centerless grinder. These parts can be lapped rapidly. When placed at the end of a line of centerless grinders, the centerless lapping machine can keep pace with the production rate of the grinders.

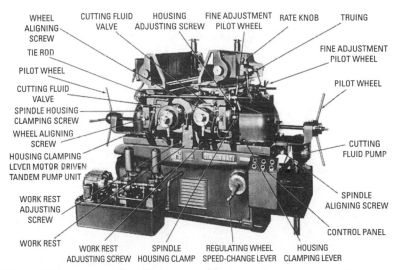

Figure 8-7 A centerless lapping machine. *(Courtesy Cincinnati Milacron Co.)*

Lapping a Cylinder

Engine cylinders that are only slightly out of round, or slightly tapered, can be trued up by lapping. Several different makes of electrical and mechanical lapping machines can be purchased for lapping engine cylinders. The operation is continued in the cylinder until a measurement check with an inside micrometer or dial gage indicates that the cylinder is true for roundness and lack of taper.

Then, the lapping compound should be washed out of the cylinder bore and the new piston rings fitted. Sometimes it is necessary to lap a new piston lightly for it to fit in the cylinder properly.

Lapping a Tapered Hole

The often-used method of finishing tapered holes with a lap that has the same taper as the hole is not recommended. The following procedure is recommended:

1. Grind the hole to size, allowing for lapping.
2. Without making any other changes, exchange the grinding wheel for a copper lap that has the same shape as the grinding wheel but a wider face.
3. Lap in the same manner that was used in grinding the hole.
4. Avoid crowding the lap.

Rotary Disc Lap

A typical example of a job for a rotary disc lap is lapping the jaws of a snap gage (Figure 8-8). The lap should be turned on the same arbor on which it is to be used, as it is practically impossible to return a lap to an arbor after it has been removed and then have it run true.

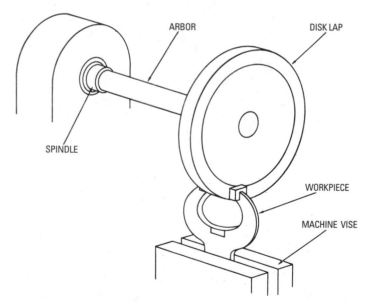

Figure 8-8 Using a rotary disc lap to lap a snap gage.

The disc lap (see Figure 8-8) is made of cast iron, and the sides are relieved, leaving only a narrow edge or flange on each side to bear against the jaws. The lap is recessed deeply to allow for truing up each time that the lap is placed on the arbor. As the table is traversed back and forth, the lap passes over the entire surface of the gage jaw, grinding it in the same manner as with a cup-shaped grinding wheel. Care should be taken not to spring the snap gage as it is clamped in the vise.

The sides of the disc lap can be trued by clamping a keen-edged cutting tool in the vise, which is similar to the truing operation on a lathe. This method is actually better than truing on the lathe because there is no change in the alignment of the lap with the work spindle. If the lap is true, a perfect contact between the lap and the gage is ensured for the entire circumference of the lap.

A charging roller (Figure 8-9) is used to charge the rotary disc lap. The abrasive is rolled in under moderate pressure. The charging

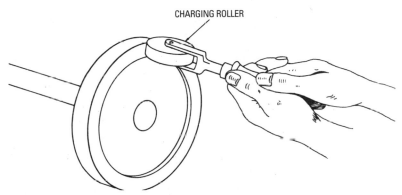

CHARGING ROLLER

Figure 8-9   A method of charging a rotary disc lap.

roller should be made of a good grade of steel, and all loose abrasive should be washed off thoroughly after the lap has been charged.

Honing

Honing is the finishing of holes with stones especially made to do the job. If a boring bar is used to make the hole or a rough drill bit makes the sides of the hole rough, then it is necessary to hone down the sides of the hole to allow a better fit with the part making contact with the sides. Honing makes a very fine finished surface. Usually, a crosshatch pattern is applied by moving the hone up and down slightly while it is rotated by a drill press or portable electric drill.

Cylinder walls on gasoline and diesel engines are honed to make sure the piston that fits inside the cylinder is able to glide up and down the cylinder wall with little friction. Figure 8-10 shows hones used for this purpose.

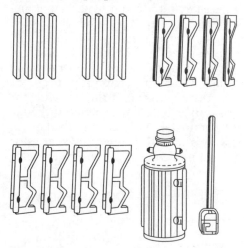

Figure 8-10 Hone set for honing cylinder walls of a small gasoline engine.

Honing is a widely accepted process apart from lapping. It is used for extremely accurate finishing of holes. A common type of honing machine has a rotating head that carries abrasive inserts for producing accurately finished holes (Figure 8-11).

 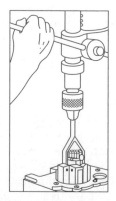

Figure 8-11 Honing using a portable hand drill and a drill press.

Lapping generally is classified as part of the grinding process. Lapping involves the use of abrasive pastes or compounds, and therefore its use is limited to the removal of very small amounts of stock. It is used in instances where a high degree of surface finish and precision are required.

Oilstones are also called hones because they are used to take away very small amounts of metal and to leave a very sharp edge. Oilstones are smooth, abrasive-type stones that are made in many shapes and sizes. They are used for sharpening, chiefly to put the finishing touches on cutting edges of tools. They should be kept clean and moist. These stones are usually kept in a box with a cover and with a few drops of clean oil left on the stone (Figure 8-12).

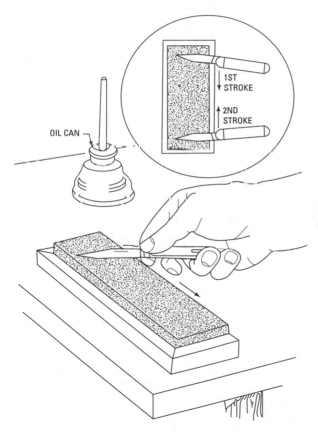

Figure 8-12 Sharpening (or honing) a knife on an oilstone.

Summary

It is often necessary to improve the finish of certain surfaces after the grinding operation has been completed. On modern tools it is often necessary to use laps and lapping powders to obtain a fine finish and improve the accuracy of the tool.

A lap is a tool used for grinding or abrading operations. There are various types of laps designed to meet the requirements of numerous grinding operations. Laps can be classified with respect to their application as either internal or external, and with respect to the material used in making the laps (such as lead, copper, brass, cast iron, and steel).

Lapping is a rubbing process for removing minute amounts of metal from surfaces that must be precision-finished with respect to dimension and smoothness. Lapping operations can be performed either by hand or by machine.

Honing is the finishing of holes with stones especially made for the job. If a boring bar is used to make the hole or a rough drill bit makes the sides of the hole rough, then it is necessary to hone down the sides of the hole to allow a better fit with the part making contact with the sides.

Oilstones are also used in honing because they are used to take away very small amounts of metal and to leave a very sharp edge. They are used primarily for finishing touches on the cutting edges of tools.

Review Questions

1. What are the four functions of lapping?
2. On what types of surfaces can lapping be used?
3. How is lapping done?
4. What is a lap?
5. How are laps classified?
6. What materials are used for laps?
7. What is considered the best lap?
8. External laps are usually made in the form of a _____.
9. What is the diameter of the various "flours" used for lapping?
10. What is the ideal abrasive for lapping?
11. Where were laps first used?
12. What is lapidary?
13. What is an important requirement of hand-lapping operations?

14. Flat laps are usually made of _____ iron.

15. List the three types of lapping medium used on lapping machines in industry.

16. Centerless lapping machines have an accuracy of finish that is within _____ to _____ microinches.

17. The sides of the disc lap can be trued by clamping a keen-edged _____ tool in the vise.

18. A typical example of a job for a rotary disc lap is lapping the _____ of a snap gage.

19. The often-used method of finished tapered holes with a lap having the same taper as the hole is not _____.

20. The disc lap is made of _____ iron.

21. What is honing?

22. What is the difference between honing and lapping?

23. Lapping is generally classified as part of the _____ process.

24. Oilstones are called _____ because they are used to take away very small amounts of metal and to leave a very sharp edge.

25. Honing usually produces a _____ pattern.

Chapter 9

Toolmaking Operations

A toolmaker is a master machinist who specializes in precision work—that is, dimensions of one-thousandth (0.001) of an inch or less. Mass production in this modern Space Age demands precision measuring. Recent developments have made the modern measuring tools more accurate and easier to read. Such things as improved finishes, wide-spaced verniers, and the expanded use of dial indicators, electronics, and optics have made measurements of a few millionths of an inch possible. Figure 9-1 shows some of these precision measuring tools in use. However, many of these precise measurements are chiefly contact measurements.

Contact measurements require a sense of touch or "feel." A skilled toolmaker can detect a difference in dimension as small as 0.00025 inch. While this "feel" varies in individuals, it can be developed with practice. The measuring tool should be properly balanced in the hand and held lightly in such a way as to bring the fingers into play in adjusting or moving the tool. Figure 9-2 shows the proper way an inside caliper is held. Note that a pair of "lock-joint transfer calipers" is being used.

Introduction

Basic or fundamental operations must be mastered by any machinist who aspires to become a master toolmaker—that is, a master of precision machining. Some of the requirements for becoming an expert toolmaker are:

- Ability to read blueprints intelligently.
- Knowledge of basic mathematics, including geometry and trigonometry.
- Ability to read and to set precision measuring tools accurately and quickly.
- Judgment as to the degree of precision required for the different tool parts.
- Inventive ability—toolmaking constantly tests the ingenuity of the toolmaker.
- Pride in keeping instruments and precision tools absolutely clean.

Figure 9-1 Precision measuring tools in use. *(Courtesy L.S. Starrett Co.)*

The art of toolmaking can best be acquired from experience. The apprentice machinist who desires to become a toolmaker practically begins a second apprenticeship in the toolroom, after completing an apprenticeship in the general shop.

At first, the apprentice is permitted to perform the more simple operations of toolmaking. As skill and experience are gained, the apprentice can be trusted with more intricate work until he or she can master any work that comes into his or her department. A good toolmaker is always attempting to improve both workmanship and knowledge of tools and toolmaking. It is only after many years of work and study that one can be considered an accomplished and expert toolmaker.

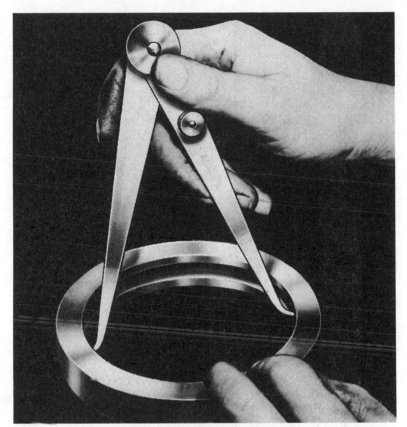

Figure 9-2   The proper way to hold an inside caliper. *(Courtesy L.S. Starrett Co.)*

A good toolmaker is always a neat worker. He or she must handle drawings carefully, lay out work painstakingly, and machine it accurately and carefully, keeping the machine clean and free from an accumulation of chips and grime. When setting up work and taking measurements, the toolmaker is careful to make sure that all bearing surfaces are clean. The tools in his or her kit are in perfect condition and adjustment because the toolmaker knows that quality workmanship is dependent upon their accuracy.

Allowances and Tolerances

The various methods of laying out and machining workpieces must be familiar to the toolmaker. The use of plugs or toolmakers' buttons must be understood thoroughly to avoid waste of time in accurately locating and boring holes. The toolmaker should possess a working knowledge of the sine bar and its application to various forms of work, and the toolmaker should be able to lay out and machine exacting angular types of work (such as dovetails, V-slides and V-ways, and plane angular surfaces).

The discussion in this chapter deals with special requirements and problems related to precision that are encountered by the toolmaker, rather than the technique used to make the various tools.

Precision Measurements

Interchangeability and duplication of parts indicate to the toolmaker that the separate parts of a mechanism must be machined and finished so closely to the dimensions that they can be assembled into a complete unit without further fitting operations. The master toolmaker is always aware of this ideal. If further fitting is necessary, the parts cannot be said to be truly interchangeable. A minute error in dimensioning can require a certain amount of fitting as the parts are assembled into a complete machine unit.

A *precision measurement* must be accurate to within one-thousandth (0.001) of an inch or fraction thereof. Such measurements are made with precision tools, which are constructed in such a manner that direct readings can be made of errors in dimension, even though the errors are minute.

Micrometers, verniers, and so on, are examples of precision tools used for direct reading of minute measurements of lengths, diameters, depths, and heights. These tools are checked for accuracy to within ten-thousandths (0.001) of an inch. Micrometers are available that measure to ten-thousandths of an inch (0.0001 inch) or to two-thousandths of a millimeter (0.002 mm). Such instruments are usually checked with gage blocks that may be accurate within a few millionths of an inch, or ±0.00003 mm, or by the use of optical flats.

Tolerance Limits

In general practice, *tolerances* are certain allowances for error (minute or otherwise) that can be tolerated. While it is possible to produce machine parts with extremely accurate measurements, extreme precision can prove too costly for some types of commercial work.

As an example of a commercial practice to avoid waste of time, labor, and money, a set of rules has been formulated to define the degree of accuracy to be expected in instances where the specifications and drawings do not call for greater precision than that for which the rules provide. The set of rules is as follows:

1. Full information regarding limits of tolerance should be clearly shown by drawings submitted, or be definitely covered by written specifications to which reference must be made by notations on the drawings.

2. Where the customer fails to supply proper data as to limits, this company's engineers will use their best judgment in deciding to what limits it is advisable to work. The company will not, in any event, assume responsibility for possible excessive cost brought about through working to closer limits than is necessary or for permitting greater latitude than may subsequently be found to be proper.

3. Where dimensions are stated in vulgar fractions with no limit of tolerance specified, it will be assumed that a considerable margin for variation from figured dimensions is available. Unless otherwise ordered, the company's engineers will proceed according to the dictates of their best judgment as to what limits should be taken.

4. For all important dimensions, decimal figures should be used and limits clearly stated on detail drawings. If decimal figures are not used for such dimensions, a notation referring to the degree of accuracy required must be placed prominently on the drawing.

5. It is frequently necessary to reduce fractions representing fourths, eighths, sixteenths, thirty-seconds, and sixty-fourths to decimal equivalents. When a dimension of this character is expressed in a decimal equivalent and carried out to three, four, or five places, and if limits are not specified, it will be assumed that a limit of ± 0.0015 is permissible, unless otherwise ordered.

6. Where dimensions are stated in decimal figures derived by other processes than those explained in paragraph 5, but with limits not specified, the following variations from dimensions stated can be expected:

   Two-place decimals: 0.005 plus or minus

   Three-place decimals: 0.0015 plus or minus

   Four-place decimals: 0.0005 plus or minus

   Five-place decimals: 0.0002 plus or minus

7. Where close dimensions, such as the location of holes from center to center in jigs, fixtures, machine parts, and other exact work of like character are required, detail drawings should be prominently marked "Accurate" and clear instructions given.

8. The dimensions of internal cylindrical gages, external ring gages, snap gages, and similar work specified to be hardened, ground, and lapped will be obtained as accurately as the best mechanical practice applying to commercial work of the particular grade specified will permit.

9. As drilled holes vary in size from 0.002 inch to 0.015 inch (and, in some instances, even more) over the size of the drill used, those required to be made accurately to definitely specified sizes should be either reamed, ground, or lapped, and detail drawings thereof should bear notations accordingly.

10. American Standard Unified form of thread and pitches will be used unless other threads are specified.

Fits and Fitting

In machine construction, many parts bear such a close and important relation to one another that a certain amount of hand fitting is essential to make the surfaces contact properly. Fits can be classified as sliding, running, forced, and shrink fits, depending on their use.

If the surfaces that are in contact move on each other, the fit is a sliding or running fit. The fit is classified as a forced or shrink fit if the surfaces make contact with enough firmness to hold them together during ordinary usage.

If two surfaces are fitted to slide on each other without lost motion laterally, the fit is called a *sliding* fit. Examples of this class of fit are the cross and transverse slides of lathes, milling machines, drilling machines, boring machines, grinding machines, and planers.

In sliding fits, the moving and stationary parts of the machines are held in contact with each other by means of adjustable contact strips or *gibs*, sometimes called packing strips. The weight of the tables of grinding and planing machines can be great enough to keep the surfaces in close contact. The running bearings of the spindles, crankshafts, line shafts, and so on, are examples of *running* fits.

In a *forced* fit, the parts are forced together under considerable pressure. Examples of this type of fit are the crankpins and axles in locomotive driving wheels and the cutterheads and spindles of numerous woodworking machines.

The amount of pressure required to force the two parts together is the limiting factor in forced fits. In "forcing" the axles into locomotive driving wheels, the specifications could limit the pressure to between 100 and 150 tons. However specified, it reduces limits to the size and requires the use of measuring tools. These measuring tools can be the direct-reading contact tools (such as the micrometer and vernier), or they can be the indirect-reading contact tools (such as the common spring caliper used in conjunction with thickness gages or *feelers*).

In a *shrink* fit, the outer or enclosing member is expanded by heating before positioning the inner member. The contraction resulting from cooling brings the parts into firm contact.

Limits of Fits

If a pin or spindle is to be forced into a hole, or if a collar, hub, flange, or other machine part is to be shrunk onto a spindle or shaft, the diameter of the spindle is usually made larger to allow for fitting rather than decreasing the size of the hole. The amount to increase the diameter of the spindle or shaft depends on the length and diameter of the hole, the metal, or material, and the shape of the hub or collar to be fitted.

The American-British-Canadian conferences held in 1952 and 1953 resulted in a proposal for a system of limits and fits. A series of standard types and classes of fits on a unilateral hole basis has been developed, so that the fit produced by mating parts in a single class will produce a similar performance throughout the entire range of sizes. This system of fits and limits was revised in 1967. Table 9-1 shows a series of standard tolerances.

Selection of the type of fit is based on the service required from the equipment being designed. The limits of size of the mating parts are established to be certain that the desired fit will be produced. Standard fits are designed by the following letter symbols:

- RC—Running or sliding fit
- LC—Locational clearance fit
- LT—Locational transition fit
- LN—Locational interference fit
- FN—Force or shrink fit

Numbers representing the class of fit are also used in conjunction with the symbols representing the type of fit—for example, RC4 represents a Class 4, running fit. Table 9-2 shows a general

Table 9-1 ANSI Standard Tolerances
(ANSI B4.1-1967, Revised 1974)

Nominal Size, in., 4 Over–To	Grade									
	4	5	6	7	8	9	10	11	12	13
	Tolerances, thousandths of an in.*									
0–0.12	0.12	0.15	0.25	0.4	0.6	1.0	1.6	2.5	4	6
0.12–0.24	0.15	0.20	0.3	0.5	0.7	1.2	1.8	3.0	5	7
0.24–0.40	0.15	0.25	0.4	0.6	0.9	1.4	2.2	3.5	6	9
0.40–0.71	0.2	0.3	0.4	0.7	1.0	1.6	2.8	*4.0	7	10
0.71–1.19	0.25	0.4	0.5	0.8	1.2	2.0	3.5	5.0	8	12
1.19–1.97	0.3	0.4	0.6	1.0	1.6	2.5	4.0	6	10	16
1.97–3.15	0.3	0.5	0.7	1.2	1.8	3.0	4.5	7	12	18
3.15–4.73	0.4	0.6	0.9	1.4	2.2	3.5	5	9	14	22
4.73–7.09	0.5	0.7	1.0	1.6	2.5	4.0	6	10	16	25
7.09–9.85	0.6	0.8	1.2	1.8	2.8	4.5	7	12	18	28
9.85–12.41	0.6	0.9	1.2	2.0	3.0	5.0	8	12	20	30
12.41–15.75	0.7	1.0	1.4	2.2	3.5	6	9	14	22	35
15.75–19.69	0.8	1.0	1.6	2.5	4	6	10	16	25	40
19.69–30.09	0.9	1.2	2.0	3	5	8	12	20	30	50
30.09–41.49	1.0	1.6	2.5	4	6	10	16	25	40	60
41.19–56.19	1.2	2.0	3	5	8	12	20	30	50	80
56.19–76.39	1.6	2.5	4	6	10	16	25	40	60	100
76.39–100.9	2.0	3	5	8	12	20	30	50	80	125
100.9–131.9	2.5	4	6	10	16	25	40	60	100	160
131.9–171.9	3	5	8	12	20	30	50	80	125	200
171.9–200	4	6	10	16	25	40	60	100	160	250

All tolerances above heavy line are in accordance with American-British-Canadian (ABC) agreements.

guide to machining processes that can be used to produce work within the indicated tolerance grades.

Tables 9-3 to 9-7 show standard tolerance limits for American Standard running and sliding fits (RC), locational fits (LC, LT, and LN), and force and shrink fits (FN).

Table 9-2   Machining Processes in Relation to Tolerance Grades

Operation	Grade
Lapping and honing	4, 5
Cylinder grinding	5, 6, 7
Surface grinding	5, 6, 7, 8
Diamond turning	5, 6, 7
Diamond boring	5, 6, 7
Broaching	5, 6, 7, 8
Reaming	6, 7, 8, 9, 10
Turning	7, 8, 9, 10, 11, 12, 13
Boring	8, 9, 10, 11, 12, 13
Milling	10, 11, 12, 13
Planing and shaping	10, 11, 12, 13
Drilling	10, 11, 12, 13

Layout

All dimensions should be given with special reference to the manner in which the tool is to be constructed. If the toolmaker must work from a certain surface in laying out the different parts of the tool and all measurements are made from that surface, the dimensions should be made to read from the surface (Figure 9-3).

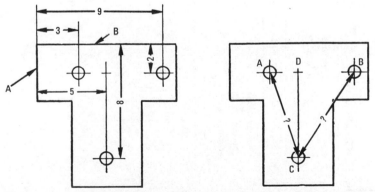

Figure 9-3 Typical dimension drawing for machine work (left), with duplicate drawing indicating additional dimensions that are necessary for precision work (right).

Table 9-3 American National Standard Running and Sliding Fits (ANSI B4.1-1967, Revised 1974)

Values shown below are in thousandths of an inch.

Nominal Size Range, in., Over–To	Class RC 1			Class RC 2			Class RC 3		
	Clearance*	Standard Tolerance Limits Hole H5	Shaft g4	Clearance*	Standard Tolerance Limits Hole H6	Shaft g5	Clearance*	Standard Tolerance Limits Hole H7	Shaft f6
0–0.12	0.1	+0.2	−0.1	0.1	+0.25	−0.1	0.3	+0.4	−0.3
	0.45	0	−0.25	0.55	0	−0.3	0.95	0	−0.55
0.12–0.24	0.15	+0.2	−0.15	0.15	+0.3	−0.15	0.4	+0.5	−0.4
	0.5	0	−0.3	0.65	0	−0.35	1.12	0	−0.7
0.24–0.40	0.2	+0.25	−0.2	0.2	+0.4	−0.2	0.5	+0.6	−0.5
	0.6	0	−0.35	0.85	0	−0.45	1.5	0	−0.9
0.40–0.71	0.25	+0.3	−0.25	0.25	+0.4	−0.25	0.6	+0.7	−0.6
	0.75	0	−0.45	0.95	0	−0.55	1.7	0	−1.0
0.71–1.19	0.3	+0.4	−0.3	0.3	+0.5	−0.3	0.8	+0.8	−0.8
	0.95	0	−0.55	1.2	0	−0.7	2.1	0	−1.3
1.19–1.97	0.4	+0.4	−0.4	0.4	+0.6	−0.4	1.0	+1.0	−1.0

	1.1	0	−0.7	1.4	0	−0.8	2.6	0	−1.6
1.97–3.15	0.4	+0.5	−0.4	0.4	+0.7	−0.4	1.2	+1.2	−1.2
	1.2	0	−0.7	1.6	0	−0.9	3.1	0	−1.9
3.15–4.73	0.5	+0.6	−0.5	0.5	+0.9	−0.5	1.4	+1.4	−1.4
	1.5	0	−0.9	2.0	0	−1.1	3.7	0	−2.3
4.73–7.09	0.6	+0.7	−0.6	0.6	+1.0	−0.6	1.6	+1.6	−1.6
	1.8	0	−1.1	2.3	0	−1.3	4.2	0	−2.6
7.09–9.85	0.6	+0.8	−0.6	0.6	+1.2	−0.6	2.0	+1.8	−2.0
	2.0	0	−1.2	2.6	0	−1.4	5.0	0	−3.2
9.85–12.41	0.8	+0.9	−0.8	0.8	+1.2	−1.8	2.5	+2.0	−2.5
	2.3	0	−1.4	2.9	0	−1.7	5.7	0	−3.7
12.41–15.75	1.0	+1.0	−1.0	1.0	+1.4	−1.0	3.0	+2.2	−3.0
	2.7	0	−1.7	3.4	0	−2.0	6.6	0	−4.4
15.75–19.69	1.2	+1.0	−1.2	1.2	+1.6	−1.2	4.0	+2.5	−4.0
	3.0	0	−2.0	3.8	0	−2.2	8.1	0	−5.6

Table 9-3 (continued)

Nominal Size Range, in., Over–To	Class RC 4 Clearance*	Class RC 4 Standard Tolerance Limits Hole H8	Class RC 4 Shaft F7	Class RC 5 Clearance*	Class RC 5 Standard Tolerance Limits Hole H8	Class RC 5 Shaft e7	Class RC 6 Clearance*	Class RC 6 Standard Tolerance Limits Hole H9	Class RC 6 Shaft e3
	Values shown below are in thousandths of an inch								
0–0.12	0.3	+0.6	−0.3	0.6	+0.6	−0.6	0.6	+1.0	−0.6
	1.3	0	−0.7	1.6	0	−1.0	2.2	0	−1.2
0.12–0.24	0.4	+0.7	−0.4	0.8	+0.7	−0.8	0.8	+1.2	−0.8
	1.6	0	−0.9	2.0	0	−1.3	2.7	0	−1.5
0.24–0.40	0.5	+0.9	−0.5	1.0	+0.9	−1.0	1.0	+1.4	−1.0
	2.0	0	−1.1	2.5	0	−1.6	3.3	0	−1.9
0.40–0.71	0.6	+1.0	−0.6	1.2	+1.0	−1.2	1.2	+1.6	−1.2
	2.3	0	−1.3	2.9	0	−1.9	3.8	0	−2.2
0.71–1.19	0.8	+1.2	−0.8	1.6	+1.2	−1.6	1.6	+2.0	−1.6
	2.8	0	−1.6	3.6	0	−2.4	4.8	0	−2.8
1.19–1.97	1.0	+1.6	−1.0	2.0	+1.6	−2.0	2.0	+2.5	−2.0

1.97–3.15	3.6	0	−2.0	4.6	0	−3.0	6.1	0	−3.6
	1.2	+1.8	−1.2	2.5	+1.8	−2.5	2.5	+3.0	−2.5
3.15–4.73	4.2	0	−2.4	5.5	0	−3.7	7.3	0	−4.3
	1.4	+2.2	−1.4	3.0	+2.2	−3.0	3.0	+3.5	−3.0
4.73–7.09	5.0	0	−2.8	6.6	0	−4.4	8.7	0	−5.2
	1.6	+2.5	−1.6	3.5	+2.5	−3.5	3.5	+4.0	−3.5
7.09–9.85	5.7	0	−3.2	7.6	0	−5.1	10.0	0	−6.0
	2.0	+2.8	−2.0	4.0	+2.8	−4.0	4.0	+4.5	−4.0
9.85–12.41	6.6	0	−3.8	8.6	0	−5.8	11.3	0	−6.8
	2.5	+3.0	−2.5	5.0	+3.0	−5.0	5.0	+5.0	−5.0
12.41–15.75	7.5	0	−4.5	10.0	0	−7.0	13.0	0	−8.0
	3.0	+3.5	−3.0	6.0	+3.5	−6.0	6.0	+6.0	−6.0
15.75–19.69	8.7	0	−5.2	11.7	0	−8.2	15.5	0	−9.5
	4.0	+4.0	−4.0	8.0	+4.0	−8.0	8.0	+6.0	−8.0
	10.5	0	−6.5	14.5	0	−10.5	18.0	0	−12.0

Table 9-3 (continued)

Nominal Size Range, in., Over–To	Class RC 7			Class RC 8			Class RC 9		
	Clearance*	Standard Tolerance Limits		Clearance*	Standard Tolerance Limits		Clearance*	Standard Tolerance Limits	
		Hole H9	Shaft d8		Hole H10	Shaft c9		Hole H11	Shaft
	Values shown below are in thousandths of an inch								
0–0.12	1.0	+1.0	−1.0	2.5	+1.6	−2.5	4.0	+2.5	−4.0
	2.6	0	−1.6	5.1	0	−3.5	8.1	0	−5.6
0.12–0.24	1.2	+1.2	−1.2	2.8	+1.8	−2.8	4.5	+3.0	−4.5
	3.1	0	−1.9	5.8	0	−4.0	9.0	0	−6.0
0.24–0.40	1.6	+1.4	−1.6	3.0	+2.2	−3.0	5.0	+3.5	−5.0
	3.9	0	−2.5	6.6	0	−4.4	10.7	0	−7.2
0.40–0.71	2.0	+1.6	−2.0	3.5	+2.8	−3.5	6.0	+4.0	−6.0
	4.6	0	−3.0	7.9	0	−5.1	12.8	0	−8.8
0.71–1.19	2.5	+2.0	−2.5	4.5	+3.5	−4.5	7.0	+5.0	−7.0
	5.7	0	−3.7	10.0	0	−6.5	15.5	0	−10.5
1.19–1.97	3.0	+2.5	−3.0	5.0	+4.0	−5.0	8.0	+6.0	−8.0

Size Range									
	7.1	0	−4.6	11.5	0	−7.5	18.0	0	−12.0
1.97–3.15	4.0	+30.	−4.0	6.0	+4.5	−6.0	9.0	+7.0	−9.0
	8.8	0	−5.8	13.5	0	−9.0	20.5	0	−13.5
3.15–4.73	5.0	+3.5	−5.0	7.0	+5.0	−7.0	10.0	+9.0	−10.0
	10.7	0	−7.2	15.5	0	−10.5	24.0	0	−15.0
4.73–7.09	6.0	+4.0	−6.0	8.0	+6.0	−8.0	12.0	+10.0	−12.0
	12.5	0	−8.5	18.0	0	−12.0	28.0	0	−18.0
7.09–9.85	7.0	+4.5	−7.0	10.0	+7.0	−10.0	15.0	+12.0	−15.0
	14.3	0	−9.8	21.5	0	−14.5	34.0	0	−22.0
9.85–12.41	8.0	+5.0	−8.0	12.0	+8.0	−12.0	18.0	+12.0	−18.0
	16.0	0	−11.0	25.0	0	−17.0	38.0	0	−26.0
12.41–15.75	10.0	+6.0	−10.0	14.0	+9.0	−14.0	22.0	+14.0	−22.0
	19.5	0	−13.5	29.0	0	−20.0	45.0	0	−31.0
15.75–19.69	12.0	+6.0	−12.0	16.0	+10.0	−16.0	25.0	+16.0	−25.0
	22.0	0	−16.0	32.0	0	−22.0	51.0	0	−35.0

All data above heavy lines are in accord with ABC agreements. Symbols H5, g4, and so on are hole and shaft designations in ABC system. Limits for sizes above 19.69 inches are also given in the ANSI Standard.

**Pairs of values shown represent minimum and maximum amounts of clearance resulting from application of standard tolerance limits.

Information Provided by Machinery's Handbook. The Industrial Press

Table 9-4 American National Standard Clearance Locational Fits (ANSI B4.1-1967, Revised 1974)

Values shown below are in thousandths of an inch

Nominal Size Range, in., Over–To	Class LC 1 Clearance*	Class LC 1 Standard Tolerance Limits Hole H6	Class LC 1 Shaft h5	Class LC 2 Clearance*	Class LC 2 Standard Tolerance Limits Hole H7	Class LC 2 Shaft h6	Class LC 3 Clearance*	Class LC 3 Standard Tolerance Limits Hole H8	Class LC 3 Shaft h7	Class LC 4 Clearance*	Class LC 4 Standard Tolerance Limits Hole H10	Class LC 4 Shaft h9
0–0.12	0 / 0.45	+0.25 / 0	0 / −0.2	0 / 0.65	+0.4 / 0	0 / −0.25	0 / 1	+0.6 / 0	0 / −0.4	0 / 2.6	+1.6 / 0	0 / −1.0
0.12–0.24	0 / 0.5	+0.3 / 0	0 / −0.2	0 / 0.8	+0.5 / 0	0 / −0.3	0 / 1.2	+0.7 / 0	0 / −0.5	0 / 3.0	+1.8 / 0	0 / −1.2
0.24–0.40	0 / 0.65	+0.4 / 0	0 / −0.25	0 / 1.0	+0.6 / 0	0 / −0.4	0 / 1.5	+0.9 / 0	0 / −0.6	0 / 3.6	+2.2 / 0	0 / −1.4
0.40–0.71	0 / 0.7	+0.4 / 0	0 / −0.3	0 / 1.1	+0.7 / 0	0 / −0.4	0 / 1.7	+1.0 / 0	0 / −0.7	0 / 4.4	+2.8 / 0	0 / −1.6
0.71–1.19	0 / 0.9	+0.5 / 0	0 / −0.4	0 / 1.3	+0.8 / 0	0 / −0.5	0 / 2	+1.2 / 0	0 / −0.8	0 / 5.5	+3.5 / 0	0 / −2.0
1.19–1.97	0 / 1.0	+0.6 / 0	0 / −0.4	0 / 1.6	+1.0 / 0	0 / −0.6	0 / 2.6	+1.6 / 0	0 / −1	0 / 6.5	+4.0 / 0	0 / −2.5
1.97–3.15	0 / 1.2	+0.7 / 0	0 / −0.5	0 / 1.9	+1.2 / 0	0 / −0.7	0 / 3	+1.8 / 0	0 / −1.2	0 / 7.5	+4.5 / 0	0 / −3.0

1.2	0	−0.5	1.9	0	−0.7	3	0	−1.2	7.5	0	−3	
3.15–4.73	+0.9	0	0	+1.4	0	0	+2.2	0	0	+5.0	0	
1.5	0	−0.6	2.3	0	−0.9	3.6	0	−1.4	8.5	0	−3.5	
4.73–7.09	+1.0	0	0	+1.6	0	0	+2.5	0	0	+6.0	0	
1.7	0	−0.7	2.6	0	−1.0	4.1	0	−1.6	10.0	0	−4	
7.09–9.85	+1.2	0	0	+1.8	0	0	+2.8	0	0	+7.0	0	
2.0	0	−0.8	3.0	0	−1.2	4.0	0	−1.8	11.5	0	−4.5	
9.85–12.41	+1.2	0	0	+2.0	0	0	+3.0	0	0	+8.0	0	
2.1	0	−0.9	3.2	0	−1.2	5	0	−2.0	13.0	0	−5	
12.41–15.75	+1.4	0	0	+2.2	0	0	+3.5	0	0	+9.0	0	
2.4	0	−1.0	3.6	0	−1.4	5.7	0	−2.2	15.0	0	−6	
15.75–19.69	+1.6	0	0	+2.5	0	0	+4	0	0	+10.0	0	
2.6	0	−1.0	4.1	0	−1.6	6.5	0	−2.5	16.0	0	−6	

Table 9-4 (continued)

Nominal Size Range, in., Over–To	Class LC 5			Class LC 6			Class LC 7			Class LC 8		
	Clearance*	Standard Tolerance Limits		Clearance*	Standard Tolerance Limits		Clearance*	Standard Tolerance Limits		Clearance*	Standard Tolerance Limits	
		Hole H7	Shaft g6		Hole H9	Shaft f8		Hole H10	Shaft e9		Hole H10	Shaft d9
Values shown below are in thousandths of an inch												
0–0.12	0.1	+0.4	−0.1	0.3	+1.0	−0.3	0.6	+1.6	−0.6	1.0	+1.6	−1.0
	0.75	0	−0.35	1.9	0	−0.9	3.2	0	−1.6	2.0	0	−2.0
0.12–0.24	0.15	+0.5	−0.15	0.4	+1.2	−0.4	0.8	+1.8	−0.8	1.2	+1.8	−1.2
	0.95	0	−0.45	2.3	0	−1.1	3.8	0	−2.0	4.2	0	−2.4
0.24–0.40	0.2	+0.6	−0.2	0.5	+1.4	−0.5	1.0	+2.2	−1.0	1.2	+2.2	−1.6
	1.2	0	−0.6	2.8	0	−1.4	4.6	0	−2.4	5.2	0	−3.0
0.40–0.71	0.25	+0.7	−0.25	0.6	+1.6	−0.6	1.2	+2.8	−1.2	2.0	+2.8	−2.0
	1.35	0	−0.65	3.2	0	−1.6	5.6	0	−2.8	6.4	0	−3.6
0.71–1.19	0.3	+0.8	−0.3	0.8	+2.0	−0.8	1.6	+3.5	−1.6	2.5	+3.5	−2.5
	1.6	0	−0.8	4.0	0	−2.0	7.1	0	−3.6	8.0	0	−4.5
1.19–1.97	0.4	+1.0	−0.4	1.0	+2.5	−1.0	2.0	+4.0	−2.0	3.6	+4.0	−3.0
	2.0	0	−1.0	5.1	0	−2.6	8.5	0	−4.5	9.5	0	−5.5
1.97–3.15	0.4	+1.2	−0.4	1.2	+3.0	−1.0	2.5	+4.5	−2.5	4.0	+4.5	−4.0
	2.3	0	−1.1	6.0	0	−3.0	10.0	0	−5.5	11.5	0	−7.0

3.15–4.73	0.5	+1.4	−0.5	1.4	+3.5	−1.4	3.0	+5.0	−3.0	5.0	+5.0	−5.0
	2.8	0	−1.4	7.1	0	−3.6	11.5	0	−6.5	13.5	0	−8.5
4.73–7.09	0.6	+1.6	−0.6	1.6	+4.0	−1.6	3.5	+6.0	−3.5	6	+6	−6
	3.2	0	−1.6	8.1	0	−4.1	13.5	0	−7.5	16	0	−10
7.09–9.85	0.6	+1.8	−0.6	2.0	+4.5	−2.0	4.0	+7.0	−4.0	7	+7	−7
	3.6	0	−1.8	9.3	0	−4.8	15.5	0	−8.5	18.5	0	−11.5
9.85–12.41	0.7	+2.0	−0.7	2.2	+5.0	−2.2	4.5	+8.0	−4.5	7	+8	−7
	3.9	0	−1.9	10.2	0	−5.2	17.5	0	−9.5	20	0	−12
12.41–15.75	0.7	+2.2	−0.7	2.5	+6.0	−2.5	5.0	+9.0	−5	8	+9	−8
	4.3	0	−2.1	12.0	0	−6.0	20.0	0	−11	23	0	−14
15.75–19.69	0.8	+2.5	−0.8	2.8	+6.0	−2.8	5.0	+10.0	−5	9	+10	−9
	4.9	0	−2.4	12.8	0	−6.8	21.0	0	−11	25	0	−15

Table 9-4 (continued)

Nominal Size Range, in., Over–To	Class LC 9			Class LC 10			Class LC 11		
	Clearance*	Standard Tolerance Limits		Clearance*	Standard Tolerance Limits		Clearance*	Standard Tolerance Limits	
		Hole H11	Shaft c10		Hole H12	Shaft		Hole H13	Shaft[CE2]
	Values shown below are in thousandths of an inch								
0–0.12	2.5	+2.5	−2.5	4	+4	−4	5	+6	−5
	6.6	0	−4.1	12	0	−8	17	0	−11
0.12–0.24	2.8	+3.0	−2.8	4.5	+5	−4.5	6	+7	−6
	7.6	0	−4.6	14.5	0	−9.5	20	0	−13
0.24–0.40	3.0	+3.5	−3.0	5	+6	−5	7	+9	−7
	8.7	0	−5.2	17	0	−11	25	0	−16
0.40–0.71	3.5	+4.0	−3.5	6	+7	−6	8	+10	−8
	10.3	0	−6.3	20	0	−13	28	0	−18
0.71–1.19	4.5	+5.0	−4.5	7	+8	−7	10	+12	−10
	13.0	0	−8.0	23	0	−15	34	0	−22
1.19–1.97	5.0	+6	−5.0	8	+10	−8	12	+16	−12
	15.0	0	−9.0	28	0	−18	44	0	−28
1.97–3.15	6.0	+7	−6.0	10	+12	−10	14	+18	−14
	17.5	0	−10.5	34	0	−22	50	0	−32

3.15–4.73	7	+9	−7	11	+14	−11	16	+22	−16
	21	0	−12	39	0	−25	60	0	−38
4.73–7.09	8	+10	−8	12	+16	−12	18	+25	−18
	24	0	−14	44	0	−28	68	0	−43
7.09–9.85	10	+12	−10	16	+18	−16	22	+28	−22
	29	0	−17	52	0	−34	78	0	−50
9.85–12.41	12	+12	−12	20	+20	−20	28	+30	−28
	32	0	−20	60	0	−40	88	0	−58
12.41–15.75	14	+14	−14	22	+22	−22	30	+35	−30
	37	0	−23	66	0	−44	100	0	−65
15.75–19.69	16	+16	−16	25	+25	−25	35	+40	−35
	42	0	−26	75	0	−50	115	0	−75

All data above heavy lines are in accordance with American–British–Canadian (ABC) agreements. Symbols H6, H7, s6, and so on are hole and shaft designations in ABC system. Limits for sizes above 19.69 inches are not covered by ABC agreements but are given in the ANSI Standard.

**Pairs of values shown represent minimum and maximum amounts of interference resulting from application of standard tolerance limits.

Table 9-5 ANSI Standard Transition Locational Fits (ANSI B4.1-1967, Revised 1974)

Nominal Size Range, in., Over–To	Class LT 1 Fit*	Standard Tolerance Limits Hole H7	Shaft js6	Class LT 2 Fit*	Standard Tolerance Limits Hole H8	Shaft js7	Class LT 3 Fit*	Standard Tolerance Limits Hole H7	Shaft k6
Values shown below are in thousandths of an inch									
0–0.12	−0.12	+0.4	+0.12	−0.2	+0.6	+0.2			
	+0.52	0	−0.12	+0.8	0	−0.2			
0.12–0.24	−0.15	+0.5	+0.15	−0.25	+0.7	+0.25			
	+0.65	0	−0.15	+0.95	0	−0.25			
0.24–0.40	−0.2	+0.6	+0.2	−0.3	+0.9	+0.3	−0.5	+0.6	+0.5
	+0.8	0	−0.2	+1.2	0	−0.3	+0.5	0	+0.1
0.40–0.71	−0.2	+0.7	+0.2	−0.35	+1.0	+0.35	−0.5	+0.7	+0.5
	+0.9	0	−0.2	+1.35	0	−0.35	+0.6	0	+0.1
0.71–1.19	−0.25	+0.8	+0.25	−0.4	+1.2	+0.4	−0.6	+0.8	+0.6
	+1.05	0	−0.25	+1.6	0	−0.4	+0.7	0	+0.1
1.19–1.97	−0.3	+1.0	+0.3	−0.5	+1.6	+0.5	−0.7	+1.0	+0.7

Range									
1.97–3.15	+1.3	0	−0.3	+2.1	0	−0.5	+0.9	0	+0.1
	−0.3	+1.2	+0.3	−0.6	+1.8	+0.6	−0.8	+1.2	+0.8
3.15–4.73	+1.5	0	−0.3	+2.4	0	−0.6	+1.1	0	+0.1
	−0.4	+1.4	+0.4	−0.7	+2.2	+0.7	−1.0	+1.4	+1.0
4.73–7.09	+1.8	0	−0.4	+2.9	0	−0.7	+1.3	0	+0.1
	−0.5	+1.6	+0.5	−0.8	+2.5	+0.8	−1.1	+1.6	+1.1
7.09–9.85	+2.1	0	−0.5	+3.3	0	−0.8	+1.5	0	+0.1
	−0.6	+1.8	+0.6	−0.9	+2.8	+0.9	−1.4	+1.8	+1.4
9.85–12.41	+2.4	0	−0.6	+3.7	0	−0.9	+1.6	0	+0.2
	−0.6	+2.0	+0.6	−1.0	+3.0	+1.0	−1.4	+2.0	+1.4
12.41–15.75	+2.6	0	−0.6	+4.0	0	−1.0	+1.8	0	+0.2
	−0.7	+2.2	+0.7	−1.0	+3.5	+1.0	−1.6	+2.2	+1.6
15.75–19.69	+2.9	0	−0.7	+4.5	0	−1.0	+2.0	0	+0.2
	−0.8	+2.5	+0.8	−1.2	+4.0	+1.2	−1.8	+2.5	+1.8
	+3.3	0	−0.8	+5.2	0	−1.2	+2.3	0	+0.2

Table 9-5 (countinued)

Over–To	Class LT 4 Fit*	Class LT 4 Standard Tolerance Limits Hole H8 H7	Class LT 4 Shaft k7	Class LT 5 Fit*	Class LT 5 Standard Tolerance Limits Hole H7	Class LT 5 Shaft n6	Class LT 6 Fit*	Class LT 6 Standard Tolerance Limits Hole H7	Class LT 6 Shaft n7
Values shown below are in thousandths of an inch									
0–0.12				−0.5	+0.4	+0.5	−0.65	+0.4	+0.65
				+0.15	0	+0.25	+0.15	0	+0.25
0.12–0.24				−0.6	+0.5	+0.6	−0.8	+0.5	+0.8
				+0.2	0	+0.3	+0.2	0	+0.3
0.24–0.40	−0.7	+0.9	+0.7	−0.8	+0.6	+0.8	−1.0	+0.6	+1.0
	+0.8	0	+0.1	+0.2	0	+0.4	+0.2	0	+0.4
0.40–0.71	−0.8	+1.0	+0.8	−0.9	+0.7	+0.9	−1.2	+0.7	+1.2
	+0.9	0	+0.1	+0.2	0	+0.5	+0.2	0	+0.5
0.71–1.19	−0.9	+1.2	+0.9	−1.1	+0.8	+1.1	−1.4	+0.8	+1.4
	+1.1	0	+0.1	+0.2	0	+0.6	+0.2	0	+0.6
1.19–1.97	−1.1	+1.6	+1.1	−1.3	+1.0	+1.3	−1.7	+1.0	+1.7

Nominal Size Range									
1.97–3.15	+1.5	0	+0.1	+0.3	0	+0.7	+0.3	0	+0.7
	−1.3	+1.8	+1.3	−1.5	+1.2	+1.5	−2.0	+1.2	+2.0
3.15–4.73	+1.7	0	+0.1	+0.4	0	+0.8	+0.4	0	+0.8
	−1.5	+2.2	+1.5	−1.9	+1.4	+1.9	−2.4	+1.4	+2.4
4.73–7.09	+2.1	0	+0.1	+0.4	0	+1.0	+0.4	0	+1.0
	−1.7	+2.5	+1.7	−2.2	+1.6	+2.2	−2.8	+1.6	+2.8
7.09–9.85	+2.4	0	+0.1	+0.4	0	+1.2	+0.4	0	+1.2
	−2.0	+2.8	+2.0	−2.6	+1.8	+2.6	−3.2	+1.8	+3.2
9.85–12.41	+2.6	0	+0.2	+0.4	0	+1.4	+0.4	0	+1.4
	−2.2	+3.0	+2.2	−2.6	+2.0	+2.6	−3.4	+2.0	+3.4
12.41–15.75	+2.8	0	+0.2	+0.6	0	+1.4	+0.6	0	+1.4
	−2.4	+3.5	+2.4	−3.0	+2.2	+3.0	−3.8	+2.2	+3.8
15.75–19.69	+3.3	0	+0.2	+0.6	0	+1.6	+0.6	0	+1.6
	−2.7	+4.0	+2.7	−3.4	+2.5	+3.4	−4.3	+2.5	+4.3
	+3.8	0	+0.2	+0.7	0	+1.8	+0.7	0	+1.8

All data above heavy lines are in accord with ABC agreements. Symbols H7, js6, and so on are hole and shaft designations in ABC system.

*Pairs of values shown represent maximum amount of interference (-) and maximum amount of clearance (+) resulting from application of standard tolerance limits.

Table 9-6 ANSI Standard Interference Locational Fits (ANSI B4.1-1967, Revised 1974)

Nominal Size Range, in., Over–To	Class LN 1 Limits of Interference* Limits	Standard Limits Hole H6	Standard Limits Shaft n5	Class LN 2 Limits of Interference* Limits	Standard Limits Hole H7	Standard Limits Shaft p6	Class LN 3 Limits of Interference* Limits	Standard Limits Hole H7	Standard Limits Shaft r6
0–0.12	0	+0.25	+0.45	0	+0.4	+0.65	0.1	+0.4	+0.75
	0.45	0	+0.25	0.65	0	+0.4	0.75	0	+0.5
0.12–0.24	0	+0.3	+0.5	0	+0.5	+0.8	0.1	+0.5	+0.9
	0.5	0	+0.3	0.8	0	+0.5	0.9	0	+0.6
0.24–0.40	0	+0.4	+0.65	0	+0.6	+1.0	0.2	+0.6	+1.2
	0.65	0	+0.4	1.0	0	+0.6	1.2	0	+0.8
0.40–0.71	0	+0.4	+0.8	0	+0.7	+1.1	0.3	+0.7	+1.4
	0.8	0	+0.4	1.1	0	+0.7	1.4	0	+1.0
0.71–1.19	0	+0.5	+1.0	0	+0.8	+1.3	0.4	+0.8	+1.7
	1.0	0	+0.5	1.3	0	+0.8	1.7	0	+1.2
1.19–1.97	0	+0.6	+1.1	0	+1.0	+1.6	0.4	+1.0	+2.0
	1.1	0	+0.6	1.6	0	+1.0	2.0	0	+1.4

Values shown below are in thousandths of an inch

1.97–3.15	0.1	+0.7	+1.3	0.2	+1.2	+2.1	0.4	+1.2	+2.3
	1.3	0	+0.8	2.1	0	+1.4	2.3	0	+1.6
3.15–4.73	0.1	+0.9	+1.6	0.2	+1.4	+2.5	0.6	+1.4	+2.9
	1.6	0	+1.0	2.5	0	+1.6	2.9	0	+2.0
4.73–7.09	0.2	+1.0	+1.9	0.2	+1.6	+2.8	0.9	+1.6	+3.4
	1.9	0	+1.2	2.8	0	+1.8	3.5	0	+2.5
7.09–9.85	0.2	+1.2	+2.2	0.2	+1.8	+3.2	1.2	+1.8	+4.2
	2.2	0	+1.4	3.2	0	+2.0	4.2	0	+3.0
9.85–12.41	0.2	+1.2	+2.3	0.2	+2.0	+3.4	1.5	+2.0	+4.7
	2.3	0	+1.4	3.4	0	+2.2	4.7	0	+3.5
12.41–15.75	0.2	+1.4	+2.6	0.3	+2.2	+3.9	2.3	+2.2	+5.9
	2.6	0	+1.6	3.9	0	+2.5	5.9	0	+4.5
15.75–19.69	0.2	+1.6	+2.8	0.3	+2.5	+4.4	2.5	+2.5	+6.6
	2.8	0	+1.8	4.4	0	+2.8	6.6	0	+5.0

All data in this table are in accordance with American-British-Canadian (ABC) agreements.

Limits for sizes above 19.69 inches are not covered by ABC agreements but are given in the ANSI Standard.

Symbols H7, p6, and so on are hole and shaft designations in ABC system.

*Pairs of values shown represent minimum and maximum amounts of interference resulting from application of standard tolerance limits.

Values shown below are in thousandths of an inch.

Nominal Size Range, in., Over–To	Class FN 1 Inter-ference*	FN 1 Standard Tolerance Limits Hole H6	FN 1 Shaft r5	Class FN 2 Inter-ference*	FN 2 Standard Tolerance Limits Hole H7	FN 2 Shaft s6	Class FN 3 Inter-ference*	FN 3 Standard Tolerance Limits Hole H7	FN 3 Shaft t6
0–0.12	0.05	+0.25	+0.5	0.2	+0.4	+0.85			
	0.5	0	+0.3	0.85	0	+0.6			
0.12–0.24	0.1	+0.3	+0.6	0.2	+0.5	+1.0			
	0.6	0	+0.4	1.0	0	+0.7			
0.24–0.40	0.1	+0.4	+0.75	0.4	+0.6	+1.4			
	0.75	0	+0.5	1.4	0	+1.0			
0.40–0.56	0.1	+0.4	+0.8	0.5	+0.7	+1.6			
	0.8	0	+0.5	1.6	0	+1.2			
0.56–0.71	0.2	+0.4	+0.9	0.5	+0.7	+1.6			
	0.9	0	+0.6	1.6	0	+1.2			
0.71–0.95	0.2	+0.5	+1.1	0.6	+0.8	+1.9			
	1.1	0	+0.7	1.9	0	+1.4			
0.95–1.19	0.3	+0.5	+1.2	0.6	+0.8	+1.9	0.8	+0.8	+2.1
	1.2	0	+0.8	1.9	0	+1.4	2.1	0	+1.6
1.19–1.58	0.3	+0.6	+1.3	0.8	+1.0	+2.4	1.0	+1.0	+2.6
	1.3	0	+0.9	2.4	0	+1.8	2.6	0	+2.0
1.58–1.97	0.4	+0.6	+1.4	0.8	+1.0	+2.4	1.2	+1.0	+2.8
	1.4	0	+1.0	2.4	0	+1.8	2.8	0	+2.2
1.97–2.56	0.6	+0.7	+1.8	0.8	+1.2	+2.7	1.3	+1.2	+3.2

	1.8	0	+1.3	2.7	0	+2.0	3.2	0	+2.5
2.56–3.15	0.7	+0.7	+1.9	1.0	+1.2	+2.9	1.8	+1.2	+3.7
	1.9	0	+1.4	2.9	0	+2.2	3.7	0	+3.0
3.15–3.94	0.9	+0.9	+2.4	1.4	+1.4	+3.7	2.1	+1.4	+4.4
	2.4	0	+1.8	3.7	0	+2.8	4.4	0	+3.5
3.94–4.73	1.1	+0.9	+2.6	1.6	+1.4	+3.9	2.6	+1.4	+4.9
	2.6	0	+2.0	3.9	0	+3.0	4.9	0	+4.0
4.73–5.52	1.2	+1.0	+2.9	1.9	+1.6	+4.5	3.4	+1.6	+6.0
	2.9	0	+2.2	4.5	0	+3.5	6.0	0	+5.0
5.52–6.30	1.5	+1.0	+3.2	2.4	+1.6	+5.0	3.4	+1.6	+6.0
	3.2	0	+2.5	5.0	0	+4.0	6.0	0	+5.0
6.30–7.09	1.8	+1.0	+3.5	2.9	+1.6	+5.5	4.4	+1.6	+7.0
	3.5	0	+2.8	5.5	0	+4.5	7.0	0	+6.0
7.09–7.88	1.8	+1.2	+3.8	3.2	+1.8	+6.2	5.2	+1.8	+8.2
	3.8	0	+3.0	6.2	0	+5.0	8.2	0	+7.9
7.88–8.86	2.3	+1.2	+4.3	3.2	+1.8	+6.2	5.2	+1.8	+8.2
	4.3	0	+3.5	6.2	0	+5.0	8.2	0	+7.0
8.86–9.85	2.3	+1.2	+4.3	4.2	+1.8	+7.2	6.2	+1.8	+9.2
	4.3	0	+3.5	7.2	0	+6.0	9.2	0	+8.0
9.85–11.03	2.8	+1.2	+4.9	4.0	+2.0	+7.2	7.0	+2.0	+10.2
	4.9	0	+4.0	7.2	0	+6.0	10.2	0	+9.0
11.03–12.41	2.8	+1.2	+4.9	5.0	+2.0	+8.2	7.0	+2.0	+10.2
	4.9	0	+4.0	8.2	0	+7.0	10.2	0	+9.0
12.41–13.98	3.1	+1.4	+5.5	5.8	+2.2	+9.4	7.8	+2.2	+11.4
	5.5	0	+4.5	9.4	0	+8.0	11.4	0	+10.0
13.98–15.75	3.6	+1.4	+6.1	5.8	+2.2	+9.4	9.8	+2.2	+13.4
	6.1	0	+5.0	9.4	0	+8.0	13.4	0	+12.0
15.75–17.72	4.4	+1.6	+7.0	6.5	+2.5	+10.6	9.5	+2.5	+13.6
	7.0	0	+6.0	10.6	0	+9.0	13.6	0	+12.0
17.72–19.69	4.4	+1.6	+7.0	7.5	+2.5	+11.6	11.5	+2.5	+15.6
	7.0	0	+6.0	11.6	0	+10.0	15.0	0	+14.0

Table 9-7 (continued)

Nominal Size Range, in. Over–To	Class FN 4			Class FN 5		
	Inter-ference*	Standard Tolerance Limits		Inter-ference*	Standard Tolerance Limits	
		Hole H7	Shaft u6		Hole H8	Shaft x7
	Values shown below are in thousandths of an inch					
0–0.12	0.3	+0.4	+0.95	0.3	+0.6	+1.3
	0.95	0	+0.7	1.3	0	+0.9
0.12–0.24	0.4	+0.5	+1.2	0.5	+0.7	+1.7
	1.2	0	+0.9	1.7	0	+1.2
0.24–0.40	0.6	+0.6	+1.6	0.5	+0.9	+2.0
	1.6	0	+1.2	2.0	0	+1.4
0.40–0.56	0.7	+0.7	+1.8	0.6	+1.0	+2.3
	1.8	0	+1.4	2.3	0	+1.6
0.56–0.71	0.7	+0.7	+1.8	0.8	+1.0	+2.5
	1.8	0	+1.4	2.5	0	+1.8
0.71–0.95	0.8	+0.8	+2.1	1.0	+1.2	+3.0
	2.1	0	+1.6	3.0	0	+2.2
0.95–1.19	1.0	+0.8	+2.3	1.3	+1.2	+3.3
	2.3	0	+1.8	3.3	0	+2.5
1.19–1.58	1.5	+1.0	+3.1	1.4	+1.6	+4.0
	3.1	0	+2.5	4.0	0	+3.0
1.58–1.97	1.8	+1.0	+3.4	2.4	+1.6	+5.0
	3.4	0	+2.8	5.0	0	+4.0
1.97–2.56	2.3	+1.2	+4.2	3.2	+1.8	+6.2
	4.2	0	+3.5	6.2	0	+5.0

Size range						
2.56–3.15	2.8	+1.2	+4.7	4.2	+1.8	+7.2
	4.7	0	+4.0	7.2	0	+6.0
3.15–3.94	3.6	+1.4	+5.9	4.8	+2.2	+8.4
	5.9	0	+5.0	8.4	0	+7.0
3.94–4.73	4.6	+1.4	+6.9	5.8	+2.2	+9.4
	6.9	0	+6.0	9.4	0	+8.0
4.73–5.52	5.4	+1.6	+8.0	7.5	+2.5	+11.6
+	8.0	0	+7.0	11.6	0	+10.0
5.52–6.30	5.4	+1.6	+8.0	9.5	+2.5	+13.6
	8.0	0	+7.0	13.6	0	+12.0
6.30–7.09	6.4	+1.6	+9.0	9.5	+2.5	+13.6
	9.0	0	+8.0	13.6	0	+12.0
7.09–7.88	7.2	+1.8	+10.2	11.2	+2.8	+15.8
	10.2	0	+9.0	15.8	0	+14.0
7.88–8.86	8.2	+1.8	+11.2	13.2	+2.8	+17.8
	10.2	0	+10.0	17.8	0	+16.0
8.86–9.85	10.2	+1.8	+13.2	13.2	+2.8	+17.8
	13.2	0	+12.0	17.8	0	+16.0
9.85–11.03	10.0	+2.0	+13.2	15.0	+3.0	+20.0
	13.2	0	+12.0	20.0	0	+18.0
11.03–12.41	12.0	+2.0	+15.2	17.0	+3.0	+22.0
	15.2	0	+14.0	22.0	0	+20.0
12.41–13.98	13.8	+2.2	+17.4	18.5	+3.5	+24.2
	17.4	0	+16.0	24.2	0	+22.0
13.98–15.75	15.8	+2.2	+19.4	21.5	+3.5	+27.2
	19.4	0	+18.0	27.2	0	+25.0
15.75–17.72	17.5	+2.5	+21.6	24.0	+4.0	+30.5
	21.6	0	+30.0	20.5	0	+28.0
17.72–19.69	19.5	+2.5	+23.6	26.0	+4.0	+32.5
	23.6	0	+22.0	32.5	0	+30.0

All data above heavy lines are in accordance with American-British-Canadian (ABC) agreements. Symbols H6, H7, s6, and so on are hole and shaft designations in ABC system. Limits for sizes above 19.69 inches are not covered by ABC agreements but are given in the ANSI Standard.

*Pairs of values shown represent minimum and maximum amounts of interference resulting from application of standard tolerance limits.

If irregularly spaced holes are to be located accurately, the dimensions should be given so that no calculations are required (see Figure 9-3). It can be noted in the illustration that all dimensions are given with respect to the two surfaces, A and B.

For most operations, the holes could be located approximately with ordinary tools. However, the distances between the holes must be given accurately for precision work, as when the toolmakers' buttons are to be used. If the distances between the holes are not given on the drawing, calculations that result in a loss of time are necessary.

For example, the three holes A, B, and C (see Figure 9-3) require that the distances between them be calculated. To determine these distances, the following calculations are necessary:

AB = 6 inches (9 – 3)
AD = 2 inches (5 – 3)
BD = 4 inches (9 – 5)
DC = 6 inches (8 – 2)

In the triangles ADC and DBC, the following calculations are necessary:

$$AC = \sqrt{(AD)^2 + (DC)^2} = \sqrt{(2)^2 + (6)^2}$$
$$= \sqrt{40} = 6.325 \text{ inches}$$

$$BC = \sqrt{(DC)^2 + (DB)^2} = \sqrt{(6)^2 + (4)^2}$$
$$= \sqrt{52} = 7.211 \text{ inches}$$

Thus, considerable time can be saved for the machinist by placing these dimensions on the drawing.

The method to be employed in laying out holes that are to be drilled depends on the accuracy desired (that is, whether the holes are to register with other holes or fixed studs). If extreme accuracy is not required, the centers for the holes can be laid out with a chalk pencil and steel rule. For locating the holes accurately (as in jig and experimental work), the centers for the holes must be laid out and scribed on the surface of the work. Scribing or layout must be done with precision tools (such as a sharp-pointed scriber, dividers, surface gage, and surface plate).

The machined surfaces should be cleaned and a copper sulfate (blue vitriol) solution applied to the surface. The treated surface, when dry, will distinctly show any lines that are made on it. Chalk can be well rubbed into the surface for the less accurate jobs.

Laying Out the Workpiece

The link (Figure 9-4) can be used here to serve as an example of a typical workpiece for precision layout. The hole centers A and B are to be made concentrically with respect to the perimeters C and D of the hubs.

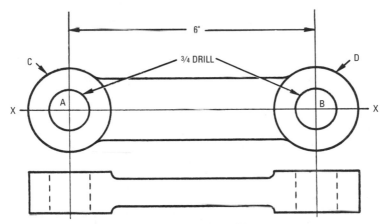

Figure 9-4 Top and side views of a link, illustrating precision necessary for layout, centering, and boring operation.

With the dividers set to a radius slightly longer than the radius of the hub, describe intersecting arcs from several points along the hub perimeter, as shown in Figure 9-5 and Figure 9-6. Here, the pairs of arcs intersect at points A, B, C, and D, forming the enclosed four-sided figure *abcd*, the center point of which is the center point of the hole that is to be drilled. If the layout were made accurately, the hole would be concentrically located with the center point of the hub.

To locate the center point, the two axes AB and CD are scribed, intersecting at center point O. The dividers can be used to check the center point for accuracy. The distances OA', OB', OC', and OD' should all be equal. If the center point should be "out" on any axis, it can be shifted until four equal measurements are obtained. The center point can then be located permanently by making a light

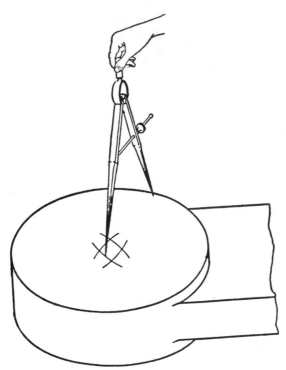

Figure 9-5 Describing intersecting arcs to locate the center point of the hub of the link so that the hole to be drilled will be concentric with the circumference of the hub.

mark with the center punch, being careful to hold the punch perpendicular to the surface.

After locating the center points of the hubs of the link, the next step is to scribe the longitudinal axis XX (see Figure 9-4). This can be done accurately by means of a surface plate and surface gage, as shown in Figure 9-7. First, set the scriber to register with the center point O, and scribe the line X across the left-hand end of the link (position A). Repeat with the same setting on the opposite end of the link (position B). The dividers (or trammels, preferably) should be set to exactly 6 inches. With one point on the center point O, an arc ab is scribed on the opposite end of the link, intersecting the axis XX at the center point O′, which is the correct center point for the right-hand end of the link (Figure 9-8). The center point O should be marked permanently with the center punch.

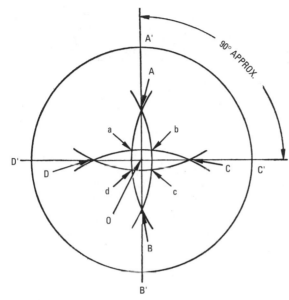

Figure 9-6 Determining the center point of the figure enclosed by the intersecting arcs to locate the center point of the hub of the link.

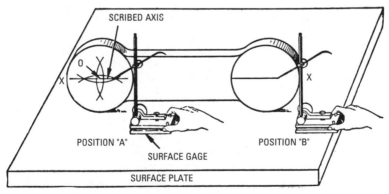

Figure 9-7 Using the surface plate and the surface gage to scribe the longitudinal axis (XX) of the link.

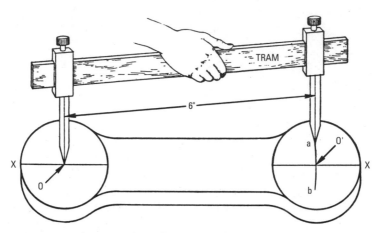

Figure 9-8 A trammel can be used to scribe the center point (O') of the opposite end of the link.

Drilling Center Holes

After the center points have been laid out, and prior to drilling, circles representing the holes are scribed with dividers. Then, the center punch marks should be enlarged to aid in starting the twist drill. Begin to drill the hole at center point O (see Figure 9-8) with the point of the twist drill in the enlarged center punch mark. The hand-feed should be used until a "dimple" is made in the workpiece. Then, the dimple should be examined to determine whether it is "on center" with the scribed circle.

If the drill point or dimple has tended to slide off to one side, a center gouge or center chisel can be used to "draw" the drill point back to the correct position, as shown in Figure 9-9. First, determine the side toward which the drill must be shifted. Then, chisel a small groove in that side (see Figure 9-9). If the dimple is very eccentric, several chiseled grooves can be required. However, if it is only slightly off center, a small groove will correct the position of the drill. Since the grooves enable the drill point to cut more easily, it is "drawn" toward the side cut by the gouge or chisel. This operation cannot be done after the full diameter of the cut has been reached.

Locating Center Points with Precision

Methods of locating center points with instruments that depend on the toolmakers' "sense of touch" for accuracy are called *contact methods*. Examples of these instruments are micrometers, vernier

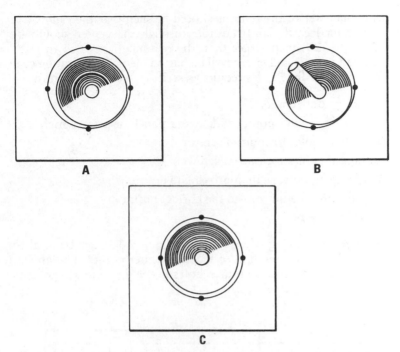

Figure 9-9   One method used to draw a drill point to bring it true with the center point of the hole, showing three positions as follows: (A) dimple is off center; (B) gouge mark on side toward which the drill point is to be drawn; and (C) dimple has returned to the center point.

calipers, dial gages, and so on. The experienced toolmaker with a finely developed sense of touch and an accurate instrument can obtain a surprising degree of accuracy by the direct-contact measurement method. Accuracy within limits of 0.0001 inch can be obtained by persons who possess a sensitive touch. Contact methods used by toolmakers to locate center points with precision include button, disc, disc and button, multidiameter discs, size blocks, and master plate.

Toolmakers' Buttons
This method of locating center points is well-adapted to small precision work. The buttons are usually 0.3, 0.4, or 0.5 inch in diameter, and they are usually ground and lapped to uniform size. The ends are finished square, or 90° to the cylindrical surface. Preferably, the

outside diameter of the buttons should be such that the radius can be determined easily, and the hole through the center should be about ⅛ (0.125) inch larger than the retaining screw, so that the button can be adjusted laterally. The button method of locating center points for drilling holes requires a series of seven operations:

1. Laying out centers
2. Marking center points with center punch or prick punch
3. Drilling holes for button screws
4. Tapping holes for button screws
5. Clamping buttons in approximate position
6. Adjusting the buttons to the correct positions
7. Boring the holes

A working drawing of a jig plate (Figure 9-10) can be used to illustrate the button method of locating center points. The dimensions necessary for locating three holes in the jig plate are given in the drawing.

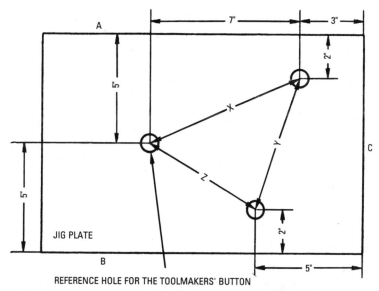

Figure 9-10 Working drawing for a jig plate. Note that the dimensions, X, Y, and Z are important in the toolmakers' button method of locating center points.

After *laying out* the center points for the holes and *marking* them lightly with a center punch, the holes for the button screws should be *drilled*. Care should be taken to select the correct size of tap drill. A tap drill should be selected that will give a snug fit, rather than the usual commercial loose fit.

Then, the holes for the button screws should be *tapped*. A typical button screw is ⅛ (0.125) inch in diameter with 40 threads per inch. Any burrs that project from the button ends after tapping should be removed carefully.

The button screws are used to clamp the buttons in an approximate position. Since the holes in the toolmakers' buttons are larger than the diameter of the button screw, the buttons are not located concentrically with the button screws, which introduces an error (Figure 9-11). A second error is introduced because the button screws have not been located with precision.

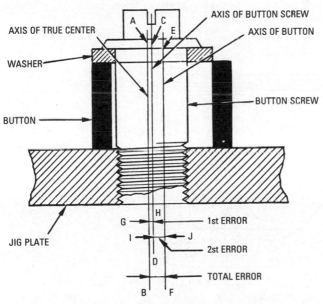

Figure 9-11 An enlarged view of a toolmakers' button, showing the errors introduced and the necessity of tapping adjustments to correct them.

Let the axis *AB* represent the true centerline, as per dimensions on the working drawing (see Figure 9-10). Regardless of the care exercised in laying out the center points on the jig plate, the intersections of the centerlines are not the precise center points of the holes to be drilled.

If we assume that the axis *AB* is the true centerline, the axis *CD* of the button screw, because of inaccuracies in layout, can be "off" a distance *GH*, which introduces the *first* error. Also, the button is not necessarily concentric with the button screw, indicated by the axis *EF*, because of clamping. It may be "off" a distance *IJ*, which introduces the *second* error, giving an accumulation of errors, or total error *BF*.

In *adjusting the buttons to their correct positions*, one button is selected as a *reference button* and is located in its correct position with respect to the jig plate by means of a micrometer, as shown in Figure 9-12. With the button screw clamped lightly, the button can be shifted minutely by rapping it lightly with a lead hammer. Micrometer measurements can be taken at each shift of the button until the button is in its correct position. Then, the button can be clamped tightly.

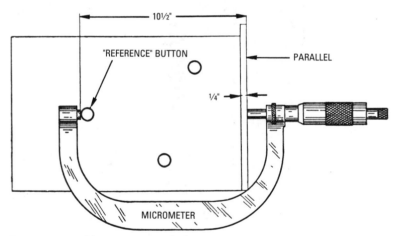

Figure 9-12 Using a micrometer and a parallel to locate the "reference" button on a jig plate.

To center two or more buttons with precision, the measurements are made with reference to the buttons themselves, as well as to the edges of the jig plate (see Figure 9-10). The center points for the buttons are dimensioned with respect to the edges *A*, *B*, and *C*. Beginning with the reference button, it is adjusted for position until its center is 10 inches (7 + 3) from the edge C and 5 inches from the edge A. In using the micrometer to adjust the button, add half the diameter of the button to the plate dimensions plus the thickness of the parallel (see Figure 9-12). Thus, for a 0.500-inch button, add 0.250 inch plus the thickness of the parallel (0.250 inch), to the plate dimension (7 inches). Therefore, to locate the reference button at 10 inches (7 + 3) from the edge C, set the micrometer at 10.500 inches (10 + 0.250 + 0.250), and adjust the button to that distance, as shown in Figure 9-12.

The next step is to adjust the button with respect to the edge A. Similarly, locate the other buttons in their correct positions. Then, clamp the reference button tightly, leaving the other two buttons clamped lightly.

The final precision adjustments involve locating the buttons in their correct positions to correspond with the measurements X, Y, and Z. In making these adjustments, the micrometer reading depends on whether an outside or an inside micrometer is used. For outside measurements, the diameter of the button (0.500 inch) must be added to the dimensions (X, Y, and Z), and subtracted if an inside measurement is used, as shown in Figure 9-13. In making these adjustments, it can be noted that the reference button (which was clamped tightly) is not moved, but the other buttons (clamped lightly) are shifted to their correct positions and clamped tightly (Figure 9-14).

After the toolmakers' buttons have been positioned correctly and clamped tightly, the final step in locating the center points of the holes for drilling or boring to size can be taken. The jig plate is then mounted on the faceplate of a lathe, with the button approximately "on center" with the axis of rotation, clamping the jig plate lightly (Figure 9-15). Rotate the faceplate by hand, checking the button with a dial indicator. Adjust the jig plate by rapping lightly with a lead hammer until the hand of the dial ceases to move, and tighten the clamping bolts. Then remove the toolmakers' button and drill or bore to the correct size.

When the workpiece is heavy and most of the jig plate is "off center," a counterweight can be attached to prevent vibration in drilling. The other toolmakers' buttons can be centered and drilled in the same manner after the first hole has been drilled.

Disc

This method is similar in principle to the button method for locating center points. The disc method is different in that the discs are made with the dimensions of the diameters such that the center

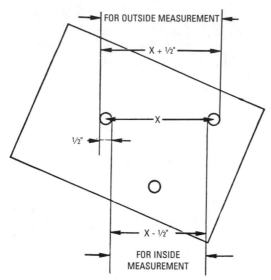

FOR OUTSIDE MEASUREMENT

X + ½"

X

½"

X - ½"

FOR INSIDE
MEASUREMENT

Figure 9-13 Diagram showing corrections necessary in setting inside and outside micrometers.

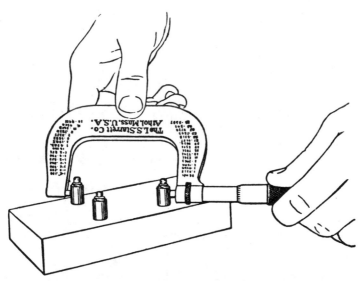

Figure 9-14 Using the micrometer to adjust the toolmakers' buttons.
(Courtesy L.S. Starrett Co.)

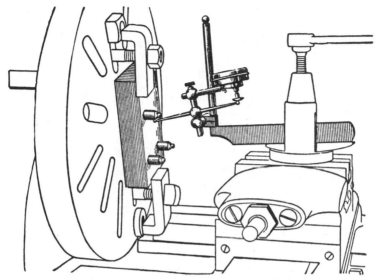

Figure 9-15 Using the dial indicator to center the toolmakers' button on the faceplate of a lathe.

points of the discs are in the correct position for drilling the holes when the circumferences of the discs are in contact with each other.

If three holes are to be drilled at distances X, Y, and Z from each other, the problem is to determine the required size of three discs (A, B, and C) so that their center points will be located at the points a, b, and c, when the circumferences of the discs are in contact (Figure 9-16). The following steps can be taken to locate the center points:

1. First, subtract the dimension Y from dimension X. This results in the distance $a'b$, which is the difference between the radii of the discs B and C.

2. Add the distance ab to the dimension Z, which will give the diameter of disc B. Describe the circumference of disc B, using the point b as the center point.

   To accomplish this, rotate the point a' around the center point b to the point b'. Then, the distance $b'c$ is equal to the diameter of disc B. You can bet the radius of disc B by bisecting the line $b'c$, using the arcs m and n. Using the line oc as the radius and the point b as center point, describe the circumference of the disc B.

3. Describe the disc C tangent to disc B.

4. Describe disc A so that it will be tangent to both disc B and disc C.

You can see in Figure 9-16 that when the three discs A, B, and C are tangent to each other, the center points of the discs coincide with the center points a, b, and c at the precise dimensions *X, Y,* and *Z.*

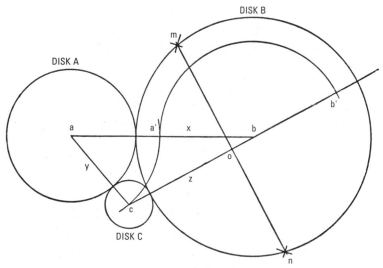

Figure 9-16   Diagram showing the disc method of locating the center points of holes by means of different-sized discs.

To locate the center points for holes that are to be drilled, the finished discs can be fastened to the workpiece in their correct locations with their circumferences in contact. As in the button method, the workpiece can be mounted on a faceplate in the lathe (see Figure 9-15) and one disc centered by means of the dial indicator applied to the center point of the disc. After the first disc has been centered, it can be removed and the hole drilled or bored. The same procedure can then be followed for the remaining holes.

Discs and Buttons
Toolmakers' buttons and discs can be used to locate center points in a combination method that combines the principles of the two

methods described previously. Each disc must have a hole at its center that fits the toolmakers' button accurately. Also, a bushing having the same diameter as that of the button and a hole in which a center punch can slide should be provided.

Figure 9-17 illustrates the disc-and-button method of locating center points of holes. In reference to the diagram, the problem is to locate the center points of six holes equally spaced on the circumference of a circle that is 6 inches in diameter.

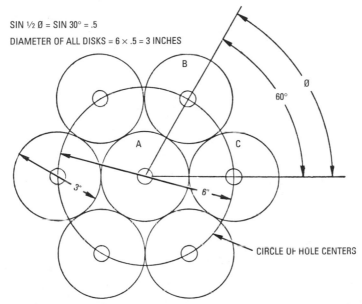

SIN ½ Ø = SIN 30° = .5
DIAMETER OF ALL DISKS = 6 × .5 = 3 INCHES

B

60°

Ø

A

C

3"

6"

CIRCLE OF HOLE CENTERS

Figure 9-17 Diagram showing the disc-and-button method of locating the center points of holes.

The size of the discs must be determined first. The size of the smaller discs can be determined by the following rule: Multiply the diameter of the circle on which the center points of the discs are located by the sine of one-half the number of degrees in the angle between the adjacent discs.

Applying the rule to the diagram (see Figure 9-17), the problem can be solved as follows:

Angle between centers of adjacent discs = 360° ÷ 6 = 60°
One-half the angle between centers = 60° ÷ 2 = 30°

Diameter of the smaller discs = $6 \times 0.5000 = 3$ inches

Diameter of the central disc = $6 - 3 = 3$ inches

Thus, in this instance, all the discs are the same size as shown in Figure 9-17. The discs and toolmakers' buttons can then be mounted on the workpiece, as explained in the previous methods.

Measuring Angles

Measuring angles between two or more lines or surfaces ("reading the angles") can be accomplished with a variety of tools, depending on the accuracy required for the job. For simple angles, a common protractor, as shown in Figure 9-18, might serve. These are graduated in degrees from 0 to 180°. The scale on this model can also be used as a depth gage.

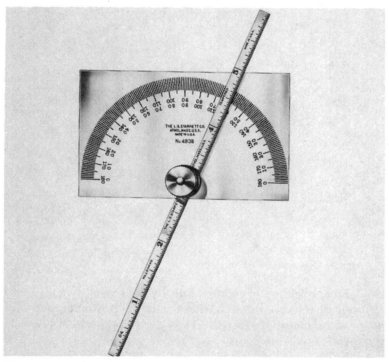

Figure 9-18 A protractor. *(Courtesy L.S. Starrett Co.)*

A bevel protractor with a vernier attachment (such as the one in Figure 9-19) can accurately measure angles to 5 minutes or one-twelfth

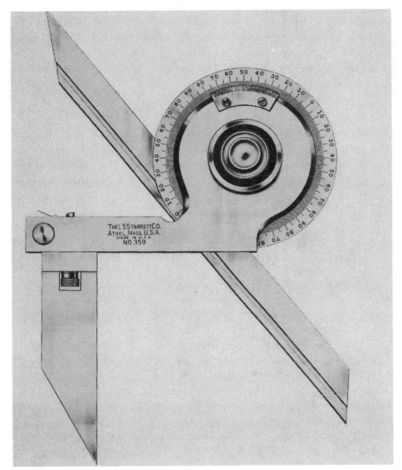

Figure 9-19 Universal bevel protractor. *(Courtesy L.S. Starrett Co.)*

of a degree. Figure 9-20 shows this bevel protractor being used to check the angle of a surface.

Multidiameter Discs for Angular Location

A line can be located at a given angle with another line by means of two discs of different diameters located with the correct distance between their center points. This method can be used to machine a surface at a given angle with another surface that is already finished.

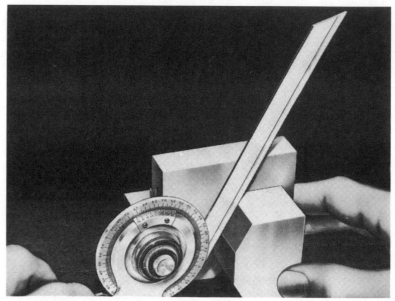

Figure 9-20 Checking the angle of a surface with a universal bevel protractor. *(Courtesy L. S. Starrett Co.)*

Figure 9-21 shows angular location by means of *nontangent discs*. In reference to the diagram, if the lower edge of the workpiece is finished and the upper edge is to be milled at an angle of 20°, the problem is to determine the proper distance *AB* to place two discs (1 inch and 2 inches in diameter) so that the upper edge of the workpiece will be tangent to both discs at the given angle. A rule can be

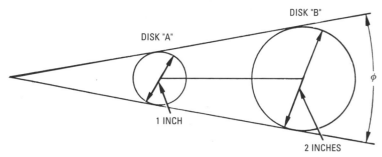

Figure 9-21 Diagram showing the angular location of a given line by means of nontangent discs.

used to solve the problem: Divide the difference between the radii of the two discs by the sine of half the required angle.

Applying the rule to Figure 9-21, the correct distance *AB* between center points of the discs can be determined as follows:

Difference between radii = 1 inch – 0.500 inch

Sine ½ Θ = sin 10° = 0.1736

Distance between center points A and B = ½ ÷ 0.1736
= 2.88 inches

Thus, the center points of the two discs A and B (see Figure 9-21) can be placed at a distance *AB* of 2.88 inches to mill the upper edge of the workpiece at an angle of 20° with the finished lower edge.

Figure 9-22 shows angular location by means of *tangent discs*. In reference to the diagram, the problem is to determine the relative sizes of two tangent discs, rather than the distance between center points, so that lines that are common tangents to the discs will be located at the desired angle.

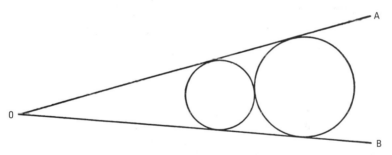

Figure 9-22 Diagram showing the angular location of a given line by means of tangent discs.

It can be noted in Figure 9-22 that the inclination of the line *OA* depends on the relative sizes of the two discs. The angle *AOB* can be made either larger or smaller by either increasing or decreasing the diameter of the larger disc. For example, if the required angle *AOB* is 20° and the diameter of the smaller disc is 1 inch, the diameter of the larger disc can be found by the application of the following three rules:

1. Multiply twice the diameter of the smaller disc by the sine of ½ the required angle (sin ½ θ or sin 10° = 0.1736):

2 × 1 × 0.1736 = 0.3473

2. Divide the product by *1* minus the sine of ½ the required angle:

$0.3473 \div (1 - 0.1736) = 0.42025$

3. Add the quotient to the diameter of the smaller disc to obtain the diameter of the larger disc:

$1 + 0.42025 = 1.42025$ inches

Thus, 1.42025 inches is the required diameter of the larger disc to form the required angle *AOB* of 20°, as shown in Figure 9-22. This method is accurate for obtaining any angle of inclination up to 40°.

Figure 9-23 shows angular location by means of a *disc and combination square*. In reference to the diagram, the problem is to determine the proper setting (distance *X*) on the blade for the head of the square to give the required angle. For example, if the required angle is 20°, and a disc with a 2-inch diameter is to be used, the setting (distance *X*) on the blade of the square can be found as follows:

1. Multiply the radius of the disc by the cotangent of ½ the desired angle—cot ½ θ or cot 10° = $\tan(90° - 10°)$ = tan 80° = 5.6713:

$1 \times 5.6713 = 5.6713$

2. Add to this product the radius of the disc:

$5.6713 + 1 = 6.6713$ inches

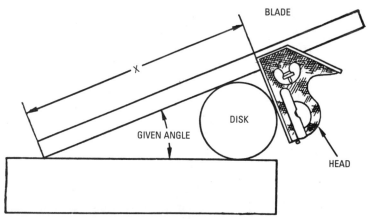

Figure 9-23 Diagram showing angular location of a given line by means of a disc and combination square.

Thus, the proper setting (distance X in Figure 9-23) on the blade of the combination square should be 6.6713 to give the required 20° angle.

Size Blocks

A *size block* is a rectangular piece that is machined to precise dimensions. Size blocks are available in various widths, so that either a single block or a combination of size blocks can be selected to locate the work properly. Size blocks provide a convenient method for locating workpieces so that holes can be drilled or bored at required center-to-center distances.

The jig plate (Figure 9-24) is an example of a workpiece on which size blocks can be used to locate the center points A and B for two holes that are to be drilled. We can assume that the jig plate is to be machined to a rectangular shape, and that the center points A and B for the two holes are located as the dimensions require.

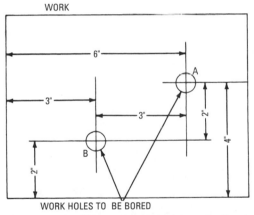

Figure 9-24 Dimensional drawing of a jig plate, used as an example of a workpiece on which size blocks can be used to locate the center points for holes that are to be drilled.

The first operation is to fasten two parallels to a faceplate of the lathe at 90° to each other, as shown in Figure 9-25. The parallels should be set off from the center of rotation at the distances given by the dimensions in the dimensional drawing (see Figure 9-24). That is, one parallel is offset 4 inches, and the other parallel is offset 6 inches. The accuracy of locating the holes depends on the precision with which the parallels are located. If the parallels are not offset precisely at 4 inches and 6 inches from center, the holes will not be correctly located, of course. Another source of error is lack of preciseness in placing the parallels at 90° to each other.

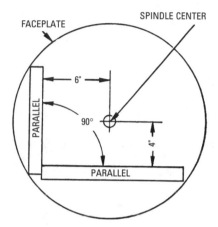

Figure 9-25 Parallels mounted on the faceplate of the lathe in accordance with dimensions given in the dimensioned drawing.

After attaching the parallels to the faceplate securely, the next step is to mount the workpiece on the faceplate so that it is in contact with both parallels, as shown in Figure 9-24. This places the workpiece correctly for drilling hole A. Then, the workpiece can be shifted to the correct position for drilling the second hole, B, by inserting size blocks between the workpiece and the parallels (see Figure 9-26).

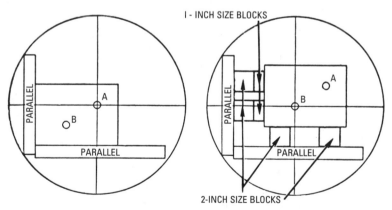

Figure 9-26 Workpiece positioned by means of parallels for drilling the first hole (left), and positioned by means of parallels and size blocks for drilling the second hole (right).

To determine the correct size blocks that can be used, note in Figure 9-26 that the center point B is located a distance of 2 inches from the edge of the workpiece. Therefore, 2-inch blocks can be used on that side. Likewise, the center point B is offset a distance of 3 inches from the center point A. Assuming that both 1-inch and 2-inch size blocks are available, but 3-inch size blocks are not available, the workpiece can be shifted a total of 3 inches from the parallel by combining the 1-inch and 2-inch size blocks (see Figure 9-26) to position the workpiece correctly for drilling the second hole, B.

Master Plate
This device is designed to be mounted on the faceplate of a lathe. Corresponding holes in the master plate and in the workpiece enable the workpiece to be shifted to correct position for drilling these holes in the workpiece. Master plates are made in any number of designs to accommodate workpieces of various shapes.

Figure 9-27 shows a dimensioned drawing of a jig plate as an example of a workpiece for which a master plate would be suitable for locating the center points for holes to be drilled or bored. The dimensioned drawing shows the location of holes that are to be drilled in the workpiece. The master plate must have corresponding holes so that the workpiece can be shifted into correct position for drilling the holes, as shown in Figure 9-28.

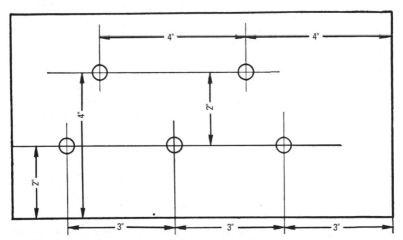

Figure 9-27   Dimensioned drawing of a jig plate, which is suitable for using as a master plate to locate the center points of holes that are to be drilled in the plate.

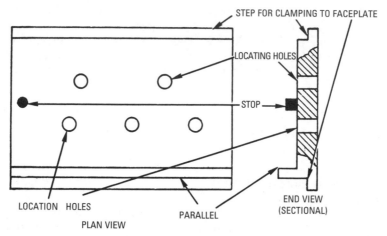

Figure 9-28 A master plate designed for positioning the jig plate.

A step is provided on each side of the master plate (see Figure 9-28) for clamping, and a projecting parallel fixes the lateral position of the work. A stop at one end of the master plate accurately positions the workpiece longitudinally. The work is positioned against both the parallel and the stop for drilling the holes. The holes are always at the correct position from the edges of the workpiece regardless of the number of duplicates made.

Clamps and tap bolts or screws that pass through the workpiece and into the master plate can be used to mount the workpiece on the master plate. If these cannot be used conveniently, solder can be used for fastening the work to the master plate.

In mounting the master plate on the faceplate of the lathe, one of the holes of the master plate is engaged with the central locating plug (Figure 9-29), and the master plate is then clamped to the faceplate. The setup is then ready for drilling or boring the hole corresponding to the hole in the master plate.

After drilling the first hole, the master plate is shifted so that first, one locating hole, and then another, can engage the central locating plug. To locate the center of a hole that is to be bored with extreme accuracy, the hole is first drilled to within 0.005 or 0.006 inch of correct size. Then it is bored to nearly finished size, a small amount of material remaining for reaming or grinding to finished size. The holes are not drilled to finished size because the point of the twist drill tends to wander off the center point.

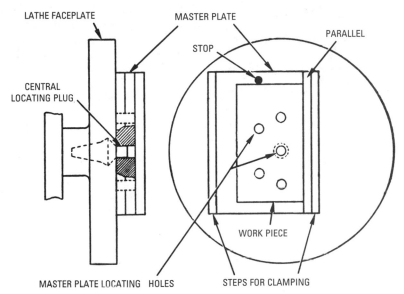

LATHE FACEPLATE

MASTER PLATE

STOP

PARALLEL

CENTRAL
LOCATING PLUG

MASTER PLATE LOCATING HOLES

STEPS FOR CLAMPING

WORK PIECE

Figure 9-29 Side view (left) showing the master plate mounted on the faceplate of the lathe, and front view (right) showing the workpiece mounted on the master plate, which is in position for drilling a hole.

Checking the Square

The four-disc method can be used to check a square for trueness of the 90° angle (Figure 9-30). Four discs of the same size are required. The accuracy of the check depends on the discs being precisely the same size.

The discs are placed in tangency with themselves and the square (see Figure 9-30). Either outside calipers or inside calipers can be used for measuring. Using outside calipers, measure the distances *AB* and *CD*. If these distances are equal, the angle between the beam and the blade of the square is true at 90°. Similarly, using inside calipers, measure the distances *EF* and *GH*. If these distances are equal, the square accurately indicates 90°.

Sine Bar for Measuring Angles

A *sine bar* is a straightedge with two attached cylindrical plugs, as shown in Figure 9-31. The instrument is used to measure angles.

Figure 9-32 illustrates the fundamental principle of the sine bar. The plugs are placed at a convenient distance (10 inches) on the straightedge. The imaginary line joining the center points of

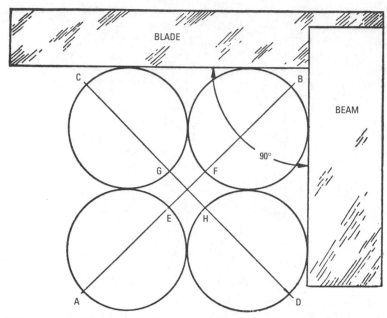

Figure 9-30 The four-disc method of checking a square for trueness of the 90° angle.

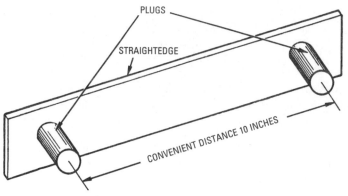

Figure 9-31 A sine bar. It is used for making angular measurements.

the two plugs represents the hypotenuse of a right triangle, as shown in Figure 9-32.

As the hypotenuse *AC* is known (10 inches), it is necessary only to measure the length of the side *BC* and divide by the length of the

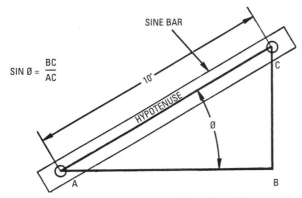

SINE BAR

$SIN \emptyset = \dfrac{BC}{AC}$

10"

HYPOTENUSE

C

Ø

A   B

Figure 9-32   Diagram showing the fundamental principle of the sine bar.

hypotenuse AC to give the sine of the angle BAC. The sine of the angle can be derived from the sine table (Table 9-8).

Note
With the common use of calculators, the angles can be taken to one-millionth of a degree. The old practice of minutes and seconds has been generally superseded by decimals.

Since it is inconvenient to measure the distance BC from the center point of the plug, a more convenient method can be used to obtain this measurement (Figure 9-33). The sine bar can be placed in position above a reference surface (such as a surface plate) and the distance A and B measured accurately with a micrometer. The distance A can be subtracted from the distance B to determine the distance (B A in Figure 9-33), which corresponds to the measurement BC in Figure 9-32.

For example, if the height $A = 2.14$ inches and $B = 7.256$ inches (see Figure 9-33), the angle of inclination of the sine bar can be determined as follows:

$$Sin \; \theta = (7.256 - 2.14) \div 10 = 0.5116$$

Then, by interpolating from the table of sines

Angle $\theta = 30°46'$ or $30.77046383°$

Figure 9-34 shows a sine bar set up to check the angle of a gage block end standard. The plugs of the sine bar are attached at the ends of the bar so one is in contact with the surface plate. The gage blocks

Table 9-8 Natural Trigonometrical Functions

Degree	Sine	Cosine	Tangent	Secant	Degree	Sine	Cosine	Tangent	Secant
0	0.00000	1.0000	0.00000	1.0000	40	0.6428	0.7660	0.8391	1.3054
1	0.01745	0.9998	0.01745	1.0001	41	0.6561	0.7547	0.8693	1.3250
2	0.03490	0.9994	0.03492	1.0006	42	0.6691	0.7431	0.9004	1.3456
3	0.05234	0.9986	0.05241	1.0014	43	0.6820	0.7314	0.9325	1.3673
4	0.06976	0.9976	0.06993	1.0024	44	0.6947	0.7193	0.9657	1.3902
5	0.08716	0.9962	0.08749	1.0038	45	0.7071	0.7071	1.0000	1.4142
6	0.10453	0.9945	0.10510	1.0055	46	0.7193	0.6947	1.0355	1.4395
7	0.12187	0.9925	0.12778	1.0075	47	0.7314	0.6820	1.0724	1.4663
8	0.1392	0.9903	0.1405	1.0098	48	0.7431	0.6691	1.1106	1.4945
9	0.1564	0.9877	0.1584	1.0125	49	0.7547	0.6561	1.1504	1.5242
10	0.1736	0.9848	0.1763	1.0154	50	0.7660	0.6428	1.1918	1.5557
11	0.1908	0.9816	0.1944	1.0187	51	0.7771	0.6293	1.2349	1.5890
12	0.2079	0.9781	0.2126	1.0223	52	0.7880	0.6157	1.2799	1.6243
13	0.2250	0.9744	0.2309	1.0263	53	0.7986	0.6018	1.3270	1.6616
14	0.2419	0.9703	0.2493	1.0306	54	0.8090	0.5878	1.3764	1.7013
15	0.2588	0.9659	0.2679	1.0353	55	0.8192	0.5736	1.4281	1.7434
16	0.2756	0.9613	0.2867	1.0403	56	0.8290	0.5592	1.4826	1.7883
17	0.2924	0.9563	0.3057	1.0457	57	0.8387	0.5446	1.5399	1.8361
18	0.3090	0.9511	0.3249	1.0515	58	0.8480	0.5299	1.6003	1.8871
19	0.3256	0.9455	0.3443	1.0576	59	0.8572	0.5150	1.6643	1.9416
20	0.3420	0.9397	0.3640	1.0642	60	0.8660	0.5000	1.7321	2.0000
21	0.3584	0.9336	0.3839	1.0711	61	0.8746	0.4848	1.8040	2.0627
22	0.3746	0.9272	0.4040	1.0785	62	0.8820	0.4695	1.8807	2.1300
23	0.3907	0.9205	0.4245	1.0864	63	0.8910	0.4540	1.9626	2.2027
24	0.4067	0.9135	0.4452	1.0940	64	0.8988	0.4384	2.0503	2.2812
25	0.7880	0.6157	1.2799	1.6243	65	0.9063	0.4226	2.1445	2.3662
26	0.4385	0.8988	0.4877	1.1126	66	0.9135	0.4067	2.2460	2.4586
27	0.4540	0.8910	0.5095	1.1223	67	0.9205	0.3907	2.3559	2.5593
28	0.4695	0.8829	0.5317	1.1326	68	0.9272	0.3746	2.4751	2.6695
29	0.4848	0.8746	0.5543	1.1433	69	0.9336	0.3584	2.6051	2.7904
30	0.5000	0.8660	0.5774	1.1547	70	0.9397	0.3420	2.7475	2.9238
31	0.5150	0.8572	0.6009	1.1666	71	0.9455	0.3256	2.9042	3.0715
32	0.5299	0.8480	0.6249	1.1792	72	0.9511	0.3090	3.0777	3.2361
33	0.5446	0.8387	0.6494	1.1924	73	0.9563	0.2924	3.2709	3.4203
34	0.5592	0.8290	0.6745	1.2062	74	0.9613	0.2756	3.4874	3.6279
35	0.5736	0.8192	0.7002	1.2208	75	0.9659	0.2588	3.7321	3.8637
36	0.5878	0.8090	0.7265	1.2361	76	0.9703	0.2419	4.0108	4.1336
37	0.6018	0.7986	0.7536	1.2521	77	0.9744	0.2250	4.3315	4.4454
38	0.6157	0.7880	0.7813	1.2690	78	0.9781	0.2079	4.7046	4.8007
39	0.6293	0.7771	0.8098	1.2867	79	0.9816	0.1908	5.1446	5.2408

Table 9-8 (continued)

Degree	Sine	Cosine	Tangent	Secant	Degree	Sine	Cosine	Tangent	Secant
80	0.9848	0.1736	5.6713	5.7588	86	0.9976	0.06976	14.3007	14.335
81	0.9877	0.1564	6.3138	6.3924	87	0.9986	0.05234	19.0811	19.107
82	0.9903	0.1392	7.1154	7.1853	88	0.9994	0.03490	28.6363	28.654
83	0.9925	0.12187	8.1443	8.2055	89	0.9998	0.01745	57.2900	57.299
84	0.9945	0.10453	9.5144	9.5668	90	1.0000	Inf.	Inf.	Inf
85	0.9962	0.08716	11.4301	11.474					

Note: For intermediate values, reduce angles from degrees, minutes, and seconds to degrees and decimal parts
of a degree (as 40° 21' 30" = 40.358°) and interpolate or consult a larger table.

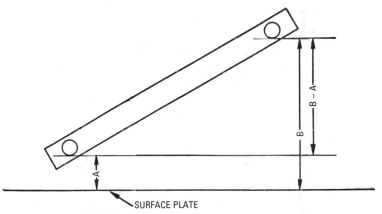

Figure 9-33 Diagram showing the use of the sine bar to obtain the measurement BC in Figure 9-32.

supporting the other plug are the exact height required. In this case, an electronic height gage is being used to check the angle. This electronic height gage is set to measure on the 0.00001-inch scale.

Sine bars can be made for special functions. Figure 9-35 shows a sine bar that has a head and tailstock to hold work between centers when checking tapers. A precision ground shaft is being held between centers, and the taper is being checked with an electronic height gage.

Sine plates (see Figure 9-36) are also used extensively in industry. They are based upon the same principle as the sine bar, but have a wider surface for holding the work to be checked. Some

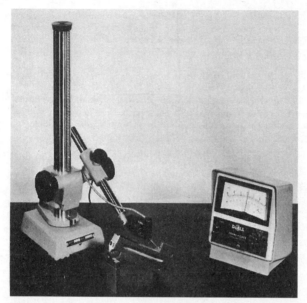

Figure 9-34 Sine bar and electronic height gage being used to check an angle. *(Courtesy DoAll Co.)*

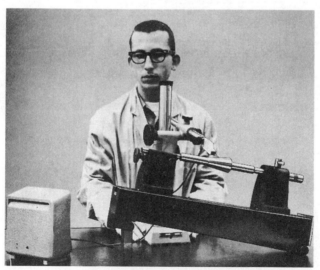

Figure 9-35 Sine bar with head and tailstock. *(Courtesy DoAll Co.)*

Figure 9-36 Sine plate in use. *(Courtesy DoAll Co.)*

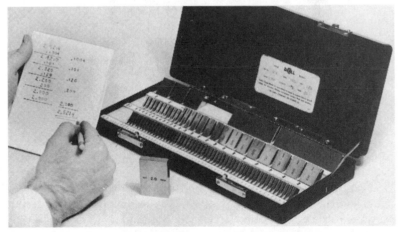

Figure 9-37 A set of gage blocks. *(Courtesy DoAll Co.)*

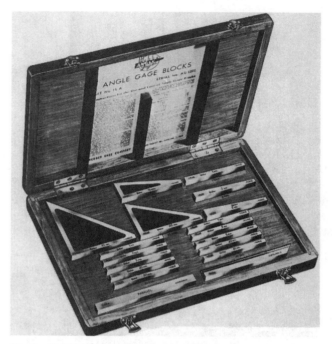

Figure 9-38   A set of angle gage blocks. *(Courtesy L.S. Starrett Co.)*

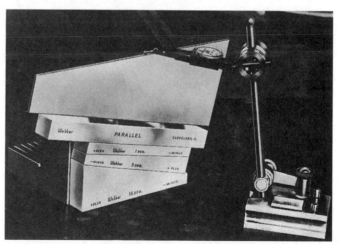

Figure 9-39   Angle gage blocks checking a 13° angle.
(Courtesy L.S. Starrett Co.)

are used to hold work while being machined, especially during grinding operations.

Note that in these illustrations of sine bars and plates, gage blocks are being used to support one end of the sine bar or plate. These are accurate within a few millionths of an inch and come in sets (such as shown in Figure 9-37) that can measure in steps as

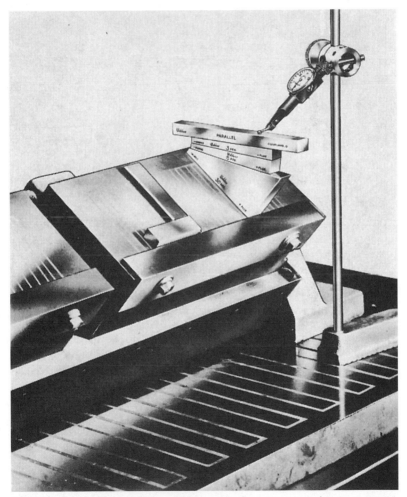

Figure 9-40 Angle gage blocks being used to set a revolving magnetic **chuck.** *(Courtesy L.S. Starrett Co.)*

small as 0.00005 inch. The illustration shows how five of these blocks can be combined to equal 2.5214 inch. Such a set can usually measure up to 12 inches.

Another method of extremely accurate angular measurement is the use of angle gage blocks. These are similar to gage blocks and provide angles as close as one-fourth of a second of a degree. A set of only 16 gage blocks (as seen in Figure 9-38) can be used to measure angles from 0° to 99° in steps of one second. These have the advantage of being assembled easily and quickly while requiring only simple addition and subtraction.

Figure 9-39 shows these blocks being used with a dial indicator to check an angle of 13°. Figure 9-40 shows these blocks being used to set a revolving magnetic chuck for a 38° angle.

The toolmaker must always keep in mind that improved measuring tools and instruments are being developed, and it is important to stay abreast of the developments in the area of precise measurement. Measuring tools are available with metric calibrations, and the toolmaker needs to be familiar with this system of measurement. The toolmaker needs to keep informed of all developments in his field, including metric measurement.

Summary

Some of the requirements for becoming an expert toolmaker are the ability to read blueprints, a knowledge of basic mathematics, the ability to read precision measuring tools, and pride in keeping instruments and precision tools clean and in good working order.

The art of toolmaking can best be acquired from experience. At first, the apprentice is permitted to perform the more simple operations of toolmaking. As skill and experience are gained, the toolmaker can be trusted with more intricate work until he or she can master any work that comes along. A good toolmaker is always a neat worker. He or she also keeps the tools in the department in perfect condition and adjustment because quality workmanship is dependent upon their accuracy.

A precision measurement must be accurate to within one-thousandth of an inch or a fraction thereof. These measurements are made with precision tools, which are constructed in such a manner that direct readings can be made of errors in dimension, even though the errors are minute.

Direct contact measurement is important, and the development of "feel" is a must. The toolmaker must understand tolerances and fits as well as proper layout. He or she must be able to "read the angles" with a variety of tools and select the proper tool for the accuracy needed. Sine bars, sine plates, gage blocks, angle gage blocks, electronic height gages, and so on, should be familiar working tools. Also, a full working knowledge of the metric system is needed.

Review Questions

1. Describe a toolmaker.
2. A skilled toolmaker can detect a difference in dimensions as small as _____ inch.
3. List five of the requirements for becoming an expert toolmaker.
4. The art of tool making can best be acquired from _____.
5. A precision measurement must be accurate to within _____ of an inch.
6. If two surfaces are fitted to slide on each other without lost motion laterally, the fit is called a _____ fit.
7. What is a forced fit?
8. What is a shrink fit?
9. What is a gib?
10. What does ANSI stand for?
11. What were the ABC agreements? What does A, B, and C stand for?
12. What is a center hole? How is it located?
13. How do you locate center points with precision?
14. What are toolmakers' buttons?
15. Explain the disc method for locating center points.
16. What is a bevel protractor used for?
17. What is a nontangent disc?
18. What is a size block?
19. What is a jig plate?
20. What is a master plate?
21. For what is a sine bar used?
22. What is an angle gage block?
23. Angle gage blocks can be used to measure angles from zero degrees to _____ degrees in steps of one second.
24. Electronic measurement tools are today very much available and highly accurate. That means greater _____ is required in the toolmaker's trade.

Chapter 10

Heat-Treating Furnaces

Various types of furnaces have been designed for heat-treating ferrous and nonferrous metals. It is difficult to classify completely the many types of furnaces used in heat-treating processes. The size and shape of the parts, the volume of production to be handled, the type of treatment needed, as well as considerations of economy and efficiency, are all factors to be considered in the selection of heat-treating furnaces.

Classification

In general, heat-treating furnaces can be classified with respect to the following:

- Source of heat, as gas-fired, oil-fired, and electric.
- Method of applying heat, as overfired, underfired, recuperative, and forced convection.
- Method of heat control, as manual or automatic.
- Protection of the work, as a controlled-atmosphere type of furnace.
- Batch or continuous operation.

Types of Furnaces

Many types of furnaces are designed for a special application or feature, such as high-speed, pusher, rotary-hearth, tilting-oven, pit, pot, box, tool, air-tempering, continuous high-temperature, and carburizing furnace. Furnaces can also be classified with respect to the maximum temperature for which the furnace was designed as: (1) low (600°F, 315.55°C); (2) medium (1200°F, 648.88°C); (3) high (1800°F, 982.22°C); and (4) extra high (2600°F, 1426.66°C). These temperature ratings are not standard ratings, but they are compiled from a large number of manufacturers' ratings for many types of furnaces. A simple production line might consist of four furnaces, as illustrated in Figure 10-1. The furnaces in order of use are preheat, high heat, quench, and draw or tempering. These furnaces would be selected according to the temperature requirements of the work being heat-treated. Other installations might include oil or water quench, soak, and/or rinse tanks.

Figure 10-1   A four-furnace production setup. *(Courtesy A.F. Holden Co.)*

Gas-Fired Oven Furnaces

A sectional view of an oven-type, heat-treating furnace is shown in Figure 10-2. A section through the burner inlet shows the basic construction of the furnace. Essentially, these furnaces have a substantial frame or casing mounted on cast-iron legs; heavy steel channels are provided to prevent sagging. Many furnaces have a counterbalanced door and lifting mechanism.

Depending on the size of the furnace, the interior walls consist of 4.5- to 5-inch fire brick backed up with insulation. The hearth is made of flat preburned forms of fire clay.

The burners are placed properly to fire underneath the working hearth without impinging on the refractory or workpiece (see Figure 10-2). This type of burner arrangement is used for heat treatment of the low-carbon steels.

For high-speed temperatures, one set of burners is placed to fire underneath the hearth, and the opposite set is placed to fire directly underneath the roof or arch of the furnace. This burner arrangement provides a rotary action for the hot products of combustion and insures rapid and uniform heating. It also prevents buildup of excessive temperatures beneath the hearth. According to the type of burner equipment provided, the operating temperatures are from 1600°F (871.11°C) to 2400°F (1315.55°C).

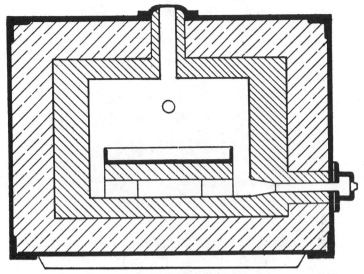

Figure 10-2: Sectional view of an oven-type, heat-treating furnace.

Electrically Heated Furnaces

Heat-treating furnaces can be heated with electricity. This is convenient, does not require a flue, is more accurately controlled, and, in certain cases, can provide higher temperatures. Electrically heated furnaces can be divided into four classes:

- *Open element*, where the heating element is exposed to the atmosphere of the furnace
- *Closed element*, with the heating element sealed into the walls and/or floor of the heating chamber
- *Immersion type*, which has a sealed resistance heating element immersed in a molten bath (usually limited to 1100°F-593.33°C)
- *Electrode type*, consisting of electrodes suspended into an electrically conductive salt

The current passes through the salt between the electrodes. The electrical currents cause the salt bath to circulate and this tends to maintain a uniform temperature.

Figure 10-3 shows an open element electric heat-treating furnace with an automatic control pyrometer. Some such furnaces are very large and have a car on which the work is loaded. Then, the car is run into the furnace through large doors.

Figure 10-3 Open element electric heat-treatment furnace.
(Courtesy Thermolyne Corp.)

Figure 10-4 shows the electrodes in a long furnace without the salt bath. Notice that the electrodes are along one side (usually the back side) and extend almost to the bottom. This furnace has four groups with four electrodes each. The number of electrodes depends on the size of the furnace, the temperature required, and the amount of work to be processed.

Pit Furnaces

Vertical furnaces of the pit type are used for heating long, slender work. These are sunk into the floor like a hole with tops that can be swung off to open the furnace. Warpage can be minimized by suspending long pieces vertically. Pit-type furnaces can be used to heat batches of small parts, which can be loaded into a basket and lowered into the furnace, as shown in Figure 10-5. The parts (bevel gears) are loaded on a rack and lowered into the deep pit furnace. This particular furnace is an atmosphere-carburizing type. Pit furnaces can be gas, oil, or electrically heated.

Figure 10-4 Electrodes of a long salt-bath furnace. *(Courtesy A. F. Holden Co.)*

Pot-Hardening Furnaces

The pot furnace is designed for indirect heating, the materials being placed in a liquid heat-transmitting medium. This is the immersion method of heat-treating small articles. The immersion method is adapted to lead hardening, liquid carburizing, liquid nitridizing, drawing, and reheating. The pot furnace has become more popular as a tool for industrial production because it is convenient, clean, accurate, economical, speedy, reduces warpage, and eliminates scale.

Pot furnaces are built for either gas, oil, or electrical firing. The basic parts of a pot furnace are the furnace, pot, drain, and hood.

The furnace itself is a circular or rectangular casing with an interior insulated lining. A cast-iron top is usually provided to support the pot. The drain is provided for accidental breakage of the pot and

Figure 10-5   Deep pit furnace being loaded with bevel gears.
(Courtesy Illinois Gear)

it is sealed by a plug to prevent passage of air into the furnace chamber while in operation. The steel hood is required when the furnace is operated with lead, cyanide, or molten salts, which give off objectionable or dangerous fumes (Figure 10-6). The hood should be provided with a flue pipe that connects it to the vent. Swinging doors make the pot accessible for charging and discharging.

When the pot furnace is either gas- or oil-fired, the burners are arranged to fire tangentially through accurately-formed combustion tunnels into the combustion chamber (Figure 10-7). This avoids flame impingement on the pot and provides through-mixing of

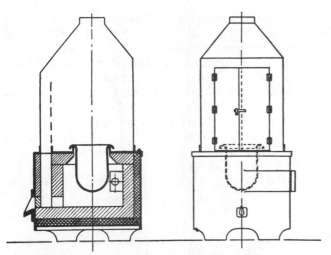

Figure 10-6   Pot furnace provided with hood for use with lead, cyanide, and other molten-bath heat-treating materials.

Figure 10-7   High-temperature, gas-fired pot furnace.
(Courtesy Johnson Gas Appliance Co.)

combustion gases. It also ensures uniform heating. Burners, vents, and holes for lighting the burners are located to provide for the desired service.

According to the requirements, either manual or automatic control can be provided for the burners. Burner capacity to

1650°F (898.88°C) can be provided for hardening service. For tempering service, temperatures as low as 350°F (176.66°C) can be obtained on the same furnace with low-pressure gas equipment by furnishing two proportional mixers, using only one for low-temperature requirements. Additions are not necessary on high-pressure gas equipment because individual burners can be shut off independently.

Industry prefers furnaces for each temperature range. This aids in maintaining the production schedule. The usual practice is to provide furnaces for the various temperature ranges required. Thus, a furnace will be used within the temperature range for which it is designed. Figure 10-8 shows a medium-temperature, gas-fired pot

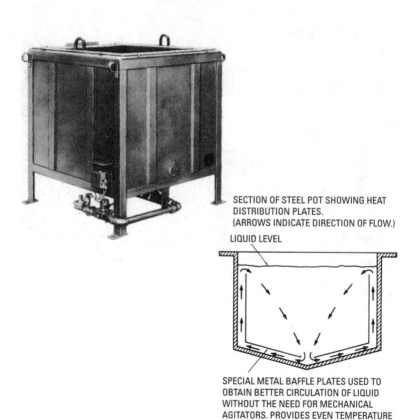

SECTION OF STEEL POT SHOWING HEAT DISTRIBUTION PLATES.
(ARROWS INDICATE DIRECTION OF FLOW.)

LIQUID LEVEL

SPECIAL METAL BAFFLE PLATES USED TO OBTAIN BETTER CIRCULATION OF LIQUID WITHOUT THE NEED FOR MECHANICAL AGITATORS. PROVIDES EVEN TEMPERATURE THROUGHOUT LIQUID.

Figure 10-8   Medium-temperature gas-fired pot furnace.
(Courtesy Johnson Gas Appliance Co.)

furnace. The section drawing shows one method of providing constant circulation within the pot. Other methods include the use of pumps and agitators.

Figure 10-9 shows the basic design of an electrically heated pot furnace. A typical heating element is formed of a single continuous helix of extra-heavy nickel chromium rod, either round or square in shape, depending on the size of the furnace. The heating element is precise in shape and spacing and is apportioned around the lining tile of the chamber so that a uniform diffusion of heat is achieved. The size and length of the coil permit a low-watt density and a lower-element temperature for the corresponding temperature of the bath.

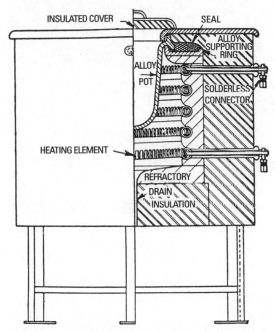

INSULATED COVER   SEAL

SEAL

ALLOY SUPPORTING RING

ALLOY POT

SOLDERLESS CONNECTOR

HEATING ELEMENT

REFRACTORY

DRAIN

INSULATION

Figure 10-9   Cutaway view of an electric pot furnace.

The terminals of the heating element are large-diameter rods of the same material brought to the outside of the furnace through special tubes with asbestos packing, which prevent heat loss and short circuits. Solderless bronze connectors are supplied for connection to the power supply.

Many of the pot furnaces in industry are the electrode type. The resistance of the bath to electrical current is used to heat and hold the bath to any desired temperature between 350°F (176.66°C) and 2350°F (1287.77°C). This is an efficient medium-size furnace that is ideal for use with conveying equipment. The control is accurate, and the maintenance cost is low. It is a practical furnace for heat-treating high-speed steels. Ceramic pots are used for the higher temperatures. Figure 10-10 shows two of these furnaces. The one on the left has a low-temperature bath.

Figure 10-10 Typical installation of two electrode-type pot furnaces.
(Courtesy A.F. Holden Co.)

Much of the increased use of the electrode-type furnace is because of the development of salt baths to use in the furnaces. Simple tests have been devised to check the baths chemically so the proper rectifiers can be added. In most cases, the workpiece can be cleaned by washing in hot water. Control of distortion is feasible and easily accomplished. The salt bath provides support of at least one-fourth of the weight of the part. A salt bath preheat will overcome machine or forming stresses. The salt baths are usually divided into three types: neutral salts, nitrate salts, and cyanide-bearing salts.

A *neutral salt bath* provides a molten heating medium inert to the workpiece. The neutral heat-treating process is designed to change only the physical characteristics of metal. It is compounded so it will not alter the chemical composition of the work and provides maximum protection for the metal during the heat-treating process.

A *nitrate salt bath* is used primarily for tempering and operates from 300°F (148.88°C) to 1100°F (593.33°C). Its function is to maintain the metal's physical property or to permit the transformation to a specific microstructure during the cool-down period. It is highly efficient in eliminating distortion and quench cracks in the workpiece. It is useful in processing finished machine parts that must be held to close tolerances.

Cyanide-bearing salt baths are used in heat-treating and case-hardening ferrous metals. They can be used to increase the carbon content of low-carbon steels to provide a hardened metal surface to resist wear and abrasion, while the soft interior gives toughness, ductility, and strength to the entire part. The depth of the case can be varied according to the needs of the part. Nitriding, a special process within this category, is producing an extremely hard case on certain types of steel at relatively low temperatures [950°F (509.99°C) to 1200°F (648.88°C)]. The process consists of placing the machined, heat-treated, and ground part in a nitrogenous medium for a preselected time at the required temperature. The steel must contain alloying elements such as aluminum, chromium, or molybdenum. The hardness of the case does not depend on quenching and will retain its hardness at temperatures up to 1200°F (648.88°C). The part can be finished after nitriding by lapping or buffing.

Recuperative Furnaces

A *recuperative furnace* is designed to recover as much heat as possible from the heated charge while it is cooling. In the common "in-and-out" or batch furnace, a cold charge is placed in the furnace, brought to temperature, and then removed to cool in the air. This method is usually too expensive for modern industry because almost 100 percent of the useful heat is wasted. Such a furnace is especially expensive to operate where slow cooling is necessary because a large quantity of additional heat is required to return the lining of the furnace to operating temperature for each successive charge.

The cost of "in-and-out" or batch-type furnace operations can be reduced slightly by placing an interchanger chamber that is large enough to hold two charges near the batch-type furnace, as shown in Figure 10-11. A heated charge from the furnace is placed in the interchanger chamber near a fresh untreated charge of material, which receives a partial preheating. Although about one-half the latent heat of the treated charge is available for preheating, the recoverable heat is actually much less because of large losses through the cold furnace walls. A 10 to 15 percent reduction in power consumption can be effected by using the interchanger chamber. However, this is not a real saving because the saving is offset by the increased handling cost.

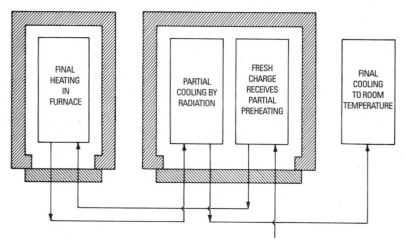

Figure 10-11 Diagram showing interchanger chamber sometimes used with a batch-type furnace to secure a slight gain in efficiency.

As shown in Figure 10-12, the interchanger chamber can be built as a single unit for a slightly improved arrangement. This arrangement can result in a 15 to 20 percent saving, which remains as only a slight improvement of the batch-type unit. The recuperative process occurs only in a single stage in this unit. Theoretically, a 50 percent saving is possible with perfect insulation and unlimited time. However, the larger portion of the heat is lost through the walls or retained in the hot charge.

The best results can be obtained by designing the furnace on the counterflow principle—to reduce the loss of heat on the outgoing material. Figure 10-13 illustrates this design. As noted in the illustration, a central heating chamber is placed between two recuperative chambers.

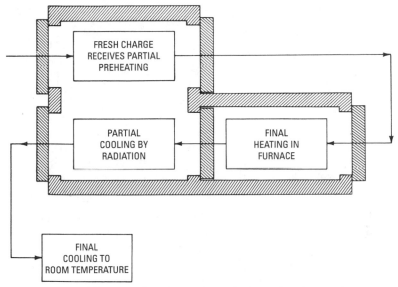

Figure 10-12 Diagram showing an interchanger chamber and furnaces built as a single unit to improve efficiency.

FINISHED CHARGES PREHEATING	FINAL HEATING	INCOMING CHARGES PREHEATED
INCOMING MATERIAL	AND SOAKING	BY FINISHED MATERIAL
RECUPERATIVE CHAMBER	HEATING CHAMBER	RECUPERATIVE CHAMBER

Figure 10-13 Diagram of a recuperative type of furnace in which heat is radiated from the finished material to the fresh stock, resulting in maximum heating efficiency.

The material being heat-treated is advanced through these chambers in opposite directions. The ingoing cold material entering the recuperative chamber is preheated by being exposed continually to warmer outgoing material. By the time the ingoing material reaches the heating chamber, it has been heated to a high degree, requiring only a small quantity of heat for a small length of time in the heating chamber to increase the temperature to the desired degree.

As the material moves outward through the second recuperative chamber, it gives up heat that is no longer needed. By continuous exposure to cooler material, it emerges from the furnace at a lower temperature.

Controlled Atmosphere

A *neutral atmosphere* is essential for the correct hardening of high-speed tools. Considerable research has been conducted with a wide range of atmospheres to determine the neutral atmosphere and develop methods of achieving it. An understanding of the changes produced by different furnace atmospheres increases the appreciation of the advantages of a truly neutral atmosphere. Three surface changes that can be produced by different furnace atmospheres are scale, decarburization, and carburization.

Scale

When air enters the heat-treating furnace, the oxygen combines with the steel to produce scale on the surface. Scale always reduces the size of the piece that is being heat-treated. Since this burning away of the metal cannot be controlled within close limits, scale is objectionable in the heat treatment of tools that should be held closely to finished size.

The quantity of scale that is formed can be reduced by introducing carbon-bearing compounds or gases. These combine with the oxygen in the air to form carbon dioxide and carbon monoxide gases in proportion to their quantities and rates of combustion. By increasing the richness of the furnace atmosphere, the quantity of scale can be reduced until a point where the scale disappears completely is reached. However, this is not a neutral point because an abrupt change causes the tools to be decarburized rather than being scaled.

Decarburization

This is a most objectionable condition because the loss of carbon causes the surface of the steel to be very soft. A furnace atmosphere that can burn the steel to form scale is also capable of combining with the carbon in the steel to produce decarburization. Therefore, it seems unusual that a scale is sometimes formed on high-speed tools, and they are not decarburized. However, the atmosphere can be slightly richer with no scale present, and the tools will become decarburized to a large degree, which is detrimental.

The burning process on the surface of the steel during formation of scale seems to form a glaze that confines the carbon. However, if the atmosphere is rich enough to prevent formation of scale, the carbon is free to escape, leaving a soft surface on the steel. This fact seems to be true for almost all high-speed tungsten steels. However, most high-speed steels that contain cobalt or molybdenum can form scale and decarburize at the same time.

After the point where the scale disappears is reached, the richness of the atmosphere can be increased to decrease the amount of decarburization until it disappears, and the neutral point, where the steel is neither giving up nor absorbing carbon, is also reached. The richness of the atmosphere can be increased further to carburize the steel and to increase the rate of carburization, depending on the characteristics of the steel, the furnace temperature, and the gases present.

Carburization

This condition is usually the least objectionable of the three conditions because a slight increase in surface hardness caused by the additional carbon is usually advantageous. The correct furnace atmosphere for hardening high-speed tungsten steel occurs at a point between decarburization and carburization—this is the only satisfactory furnace atmosphere for hardening high-speed cobalt and molybdenum steels. The time element is not critical. Tools can be given an ample "soaking" period to assure maximum hardness.

Tools can be heated in a controlled atmosphere for more than an hour at 2350°F (1287.77°C) without any change in size or condition of the surface to indicate the application of excessive heat. The neutral furnace atmosphere envelopes or surrounds the piece that is being heat-treated to protect it.

A precisely controlled atmosphere, maintained in sufficient volume to fill the furnace chamber completely, protects both the surface and the structure of the steel, preventing scale, decarburizing, pitting, and other harmful effects. This, in turn, reduces grinding and finishing costs, improves structure and production life in service, and eliminates spoilage caused by various forms of harmful effects, thereby accomplishing overall savings in tool cost.

Controlled-Atmosphere Furnaces

Various types of controlled-atmosphere furnaces are designed to fulfill the special heat-treating requirements in shops where tools and dies are heat-treated in large volumes. These furnaces are also adaptable to shops in which the volume is not large but the tools and dies are of a design that substantial loss could result from inadequate furnace equipment. Figure 10-14 shows a diagram of a controlled-atmosphere furnace.

An air-gas mixture is precombusted in the precombustion chamber (*A* in Figure 10-14) to form a stable protective atmosphere of

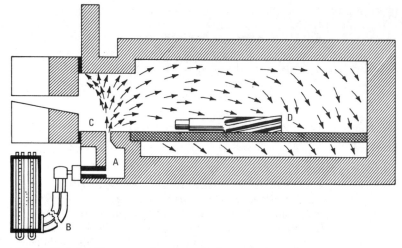

Figure 10-14 Diagram of a controlled-atmosphere furnace.

any desired proportion to fulfill the requirement of a particular type of steel that is being heat-treated. A manometer *B* regulates the air-gas mixture entering the precombustion chamber. The precombusted atmospheric mixture is forced upward through an unbroken slot *C* across the furnace opening with sufficient pressure to seal the opening positively against the entrance of outside air and with sufficient volume to fill the heating chamber—surrounding the workpiece *D* at all times with a protective atmosphere suitable for the makeup or composition of the steel in the workpiece that is being heat-treated.

This method introduces the principle of precombustion of a gas-air mixture and the use of the resulting products of combustion to provide a protective atmospheric medium in the heat-treating furnace. To obtain the controlled atmosphere:

- The door of the furnace is sealed against the entrance of the outside atmosphere.
- An enveloping or furnace atmosphere of any analysis that is desired is provided for protection of the work undergoing heat treatment.

The controlled atmosphere can be varied by turning the valves provided for that purpose to give the desired manometer settings. Once the most effective atmosphere has been determined, it can be duplicated accurately in the future by repeating the settings on the manometer. A controlled atmosphere was needed in electric furnaces where gas was not used. This led to the use of several inert gases to provide the controlled atmosphere, which has been so successful that they are being used in gas-heated furnaces. It is rather easy if the heating chamber for the work is isolated from the burners. The savings on machining costs after heat-treating make it economical to operate. Figure 10-15 shows atmosphere-controlled carburizing furnaces in use.

Figure 10-15 Atmosphere-controlled carburized furnaces.
(Courtesy Illinois Gear)

Temperature Control of Heat-Treating Furnaces

Regulation of the temperature of the workpiece being treated is the basic factor in modern industrial furnace control. The controls must regulate automatically. This must be done at the lowest possible cost in spite of changes in production, ambient temperature, supply voltage, fuel characteristics, or other changes in the heat-treating process.

It is the temperature of the workpiece that is important—not the furnace temperature. The furnace is a basic component of the temperature-regulating system. The temperature of the work is controlled, either directly or indirectly, by regulating the furnace temperature.

In addition to the furnace, or heating equipment, temperature-control systems include the following parts (Figure 10-16):

- The primary element or sensing device
- The instrument or controller
- The final control element, or power-control device

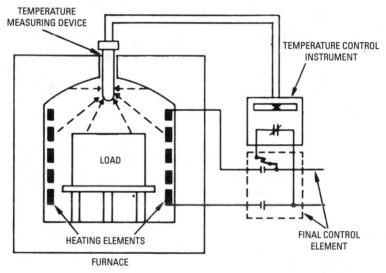

Figure 10-16 The heating elements of a heat-treating furnace are controlled by relays and switches. Temperature-sensing devices cause the relays to energize and thereby turn the power on or off. The types of heat that affect the temperature control are indicated by the arrows.

Most temperature controls for heat-treating furnaces are referred to (incorrectly) as pyrometers. A pyrometer is a device for measuring the degrees of heat higher than those recorded by a mercurial thermometer. It includes the sensing device and a meter to indicate the number of degrees.

Results Are Important

The results of the temperature control, rather than the absolute temperature, are more important in measuring a controlled temperature. Correct operating temperature is determined by trial-and-error and by the results of heat-treating—regardless of the temperature indicated by the measuring device.

Absolute temperature is difficult to determine because:

- Temperature is not constant in most furnaces.

- The temperature in the various parts of the furnace or workpiece is always different because of differences in losses and distribution of heat throughout the furnace.

- The temperature-measuring device indicates the furnace temperature at a single point in the furnace rather than the temperature of the workpiece.

Response

The temperature-measuring device continually absorbs or loses heat because furnace temperature changes continually (Figure 10-17). Accuracy of the device in following the changes depends on its speed in responding. In general, the temperature-measuring device should respond a little more rapidly than the workpiece that is being heat-treated.

A measuring device with quick response should be used when thin steel strip is heated. A more sluggish device can be used when heating heavy castings or tightly wound coils. Measuring devices respond more rapidly at higher temperatures.

Measuring Temperature

The heat energy in a body is indicated by temperature. Temperature rarely can be measured directly, but it is inferred by observation of its effect on a known substance.

All materials are affected by temperature, and an almost unlimited number of methods are used to measure it. The most familiar method is the bimetal thermostat commonly used in home furnace control. It is unsatisfactory for remote indication, and it has limited maximum temperature [around 1100°F (593.33°C)]. A nonindicating unit [up to 1800°F (982.22°C)] can be used for overtemperature protection.

Bulb-Type Devices

This type of detector uses a metal bulb filled with liquid, liquid and vapor, or gas connected to the operating mechanism through a thin

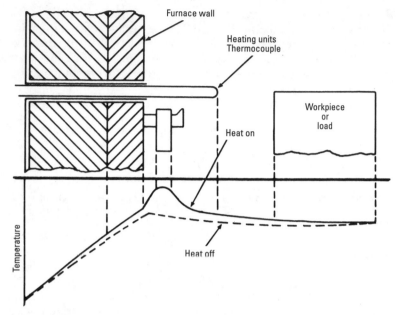

Figure 10-17 Variations in temperature inside a furnace during the on and off cycles. Note the temperature is greatest next to the heat source; then it tapers off, there being less variation near the workpiece (load) than through the brickwork or insulation.

capillary tube. The expansion of the material, transmitted through the capillary tube, operates a bellows, diaphragm, or helix, which, in turn, operates the control mechanism and sometimes an indicating or recording device. The bulbs can be operated at a range of −300°F (−184°C) to +1200°F (648.88°C), depending on the material used in the bulb.

Bulb-type devices are inexpensive, accurate within reason, and commonly used for control applications such as the cooling of water. Higher temperatures are limited chiefly because of the capillary tubing. Therefore, the bulbs are not used to a great extent for furnace temperature control.

Temperature-Sensitive Devices

Temperature-sensitive resistors are becoming more common for the higher-temperature ranges. They consist of a small coil of copper, nickel, or platinum wire wound on an insulating form and enclosed by a protective tube. The change in resistance of these materials is uniform and can be predicted for changes in temperature. These devices are among the most accurate that are available if they are used properly. They are particularly adapted to low temperatures or to small differences in temperature. They are very accurate for temperatures below 1600°F (871.11°C), but should not be used continuously for higher temperatures. The maximum temperature is about 3200°F (1760°C). The chief disadvantage of temperature-sensitive resistors in industrial furnace work is the fact that most materials are adversely affected by the common reducing atmospheres. These are usually referred to as *resistance pyrometers*.

Radiation Detectors

The basic components of *radiation detectors* are the small thermocouples connected in series with their hot junctions close together. A mirror or lens is used to focus radiant energy on the hot junctions. Radiation detectors can be used at temperatures above that usually permitted for standard thermocouples or other devices used to measure the temperature of moving objects, such as steel strips.

The radiation detector's usual temperature range is 800°F (426.66°C) to 3300°F (1815.55°C), but it can be used at temperatures as low as 125°F (51.66°C) and as high as 6000°F (3315.55°C). Low temperatures require a special control for the reference junction. They are difficult to calibrate in terms of absolute temperature. The practical temperature span and usable range is limited because radiation detectors are nonlinear. They should be used only in the upper half of their calibrated range.

Metal or ceramic tubes of either the open-end or closed-end type are used to protect radiation detectors. The instruments are usually furnished with calibrating rheostats because open-end tubes may have to be calibrated for emittance from the hot body.

Temperature measurements should not be taken at face value. *Radiant energy* that is given off by a hot surface is a function of the surface temperature and surface emittance. These characteristics vary for different materials and even for the same material at different temperatures.

Several problems are involved in measuring the load temperature in a closed furnace. If the load temperature is constant, the radiant energy absorbed must equal that which it emits. If the load is opaque, all the radiant energy falling on it must be absorbed and subsequently emitted, or reflected.

Radiation detectors cannot differentiate between radiated heat and reflected heat (Figure 10-18). If the temperature of the workpiece is at the same temperature as the furnace walls, it will show black body characteristics, and the temperature reading will be accurate. For the same reason, an accurate furnace wall temperature can be obtained if the volume of work is small compared with the volume of the furnace.

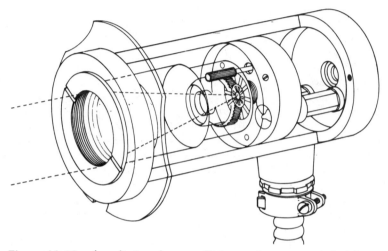

Figure 10-18   A radiation detector. This type is recommended for furnaces operating at 2700°F (1482.22°C) or above. The radiation detector cannot differentiate between radiant heat and reflected heat.

If the temperature of the load is lower than the temperature of the furnace walls, the radiant heat reflected by the work represents a hotter temperature than that of the work, and the temperature reading will be higher than that of the work. Similarly, if the hot workpiece is surrounded by a temperature lower than its own temperature, the temperature measured will be lower. If the temperature of the furnace wall and emittance of the hot object are known, the error can be calculated. This is difficult to calculate with a high degree of accuracy.

The radiation detector is practically the only satisfactory commercial device for furnaces operated at temperatures above 2700°F (1482.22°C). It is most satisfactory for moving objects or objects that cannot be touched. It is sometimes used to measure the temperature of workpieces heated by induction. For this application, radiation detectors that require a small viewing area and respond quickly (98 percent in 1 second) are available.

Consideration must be given to the effect of discontinuity in measuring and controlling the temperature of moving objects that do not move continuously. The measuring device must be made slow (damped) purposely to prevent a dip in temperature indication each time there is discontinuity. This dip in temperature can cause undesirable control action.

In addition to interpretation difficulties, radiation detectors have other disadvantages in that they are more expensive than most other temperature-measuring devices. The devices are also subject to calibration errors because of errors caused by fogging, changes of emittance, and the effects of gases, smoke, and flames.

These radiation pyrometers are not connected to the furnace. No part of the instrument is inserted in the high heat to be measured. The tube is pointed toward the open door of the furnace or at an opening, and the temperature can be read on the indicator. They can be hand held or mounted on a tripod.

Optical Pyrometers
There are several types of *optical pyrometers*. In one type, an electric lamp filament is heated to the same color as that of the part being heated, the temperature of which is needed. The current consumed is indicated, and the corresponding temperature is determined. There are several other types of optical pyrometers. These pyrometers are used to estimate the highest temperatures and may be used for temperatures above 3000°F (1648.88°C).

Thermocouples
The *thermocouple* is simple, low in cost, and adaptable. It is the most common and most versatile temperature-measuring and controlling device. The thermocouple is part of the most commonly-used pyrometer known as the *thermoelectric pyrometer*. The thermocouple is placed in the furnace and connected by wires to a meter, which may or may not be close to the furnace. Proper selection of thermocouples and their uses is complicated. A vast amount of literature has been written on these subjects.

Basic Principle
Basically, a thermocouple consists of two wires joined together at one end (hot junction) and connected at the other end to an electric measuring device. When the junctions of the metals are maintained at different temperatures, electricity is produced. A voltage or

electromotive force (emf) is produced if the circuit is open, and a current flows if the circuit is closed.

If one of the junctions is kept at a fixed temperature, the emf will give, for any two given metals, a fixed relation to the temperature at the other junction. This combination is known as a thermocouple, or a "couple," and can be used as a temperature-measuring and controlling device.

The emf and the current produced by the thermocouples are an approximate linear function of the temperature difference between the hot and the cold junctions. Calibration tables are based on a cold junction temperature of 32°F (0°C). The cold junctions are often placed in a thermos bottle of ice water. Most temperature units use room temperature for the cold junction. A compensating resistor or other means is provided for calibration.

The two wires are insulated by porcelain beads in a typical thermocouple (Figure 10-19). A junction block is provided for the thermocouple and the extension lead wire that connects it to the temperature-measuring instrument. The junction block and the thermocouple are enclosed and held by a head, usually a diecasting.

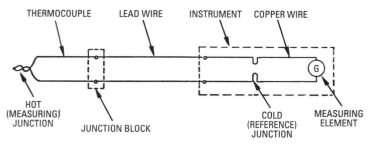

Figure 10-19   Diagram of a typical electrical circuit for a thermocouple.

Protection of Thermocouple
Thermocouples are usually enclosed in the modern furnaces to protect them from gases, liquids, and mechanical damage, although open thermocouples are sometimes used. The enclosed thermocouples are supported, and installation in pressure or liquid vessels is simplified.

The location of the thermocouple is important. Several points should be considered:

- The thermocouple should be affected only by the heating unit that it controls. Except for protection against over-temperature, it should never be placed between two furnace zones.

• The thermocouple should be inserted in the furnace far enough to prevent cooling by conduction, but not far enough for the thermocouple to droop. Its calibration should not be affected by a temperature gradient throughout its length in the furnace.

• Thermocouples mounted horizontally in well-insulated furnaces are inserted beyond the inside brickwork a distance of four to six times their outside diameter. Immersion up to ten times the outside diameter may be required for some applications. The depth of immersion should not be changed once the thermocouple is installed.

• The thermocouple should not be placed near a heating element. Similarly, the thermocouple should be protected from direct contact with the flames in open-fired gas or oil furnaces.

• When thermocouples are placed in protective atmospheres, the products of combustion should not be allowed to escape past the connection block. Escaping gases can heat up the connections to the lead wire, causing errors in temperature readings. However, they can also fill the conduits with explosive gases. The head should be screwed to a pipe nipple that is welded to the furnace casing or otherwise sealed against the loss of furnace atmosphere.

• The lead wires are color coded, and they should be connected to the correct poles. If the instrument reads in the wrong direction, the connections should be reversed at the connection block rather than at the instrument.

Heat Radiation

Virtually all the heat received by the thermocouple is by radiation, at temperatures above 1000°F (537.77°C). In this range, the thermocouple is a radiation detector. Many of the aforementioned observations on radiation apply to detectors.

Calibration

When great accuracy is required, thermocouples can be sent to the National Bureau of Standards in Washington, D.C., for calibration. Most thermocouples are guaranteed for ¾ of 1 percent of the specified calibration at furnace temperatures above 500°F (260°C). In some instances, thermocouples made of special-grade wire provide about one-half the standard limit of error.

Many combinations of materials have been developed for thermocouples. Among these materials are copper-constantan, chromel-constantan, steel-alumel, nickel-tungsten, and others. The majority of thermocouples used in industrial furnace work are shown in Table 10-1.

Table 10-1 Basic Thermocouple Types

Type	Temperature	Uses
Iron-constantan	30–1400	Low temperatures; air drawing; copper or brass annealing; ovens
Chromel-alumel	0–2100	Copper annealing; wire enameling; air atmosphere; medium-temperature heat-treating
Nickel-nickel (18% molybdenum)	to 2150	Copper brazing; heat-treating steel; and stainless
Platinum-Platinum-rhodium	to 2800	Forging furnaces; high-temperature heat-treating; laboratory work

Lead Wire

Expensive lead wire connecting the thermocouple to the temperature instrument is not necessary. Extension lead wires of the same or more inexpensive materials can be used. These are usually two insulated 16-gage wires in a common outer braid. If it is possible, there should be no joints between the thermocouple and the instrument.

If a joint is necessary, the wires should be scraped clean, spliced, and soldered or brazed. The extension wires should be run in grounded conduits for protection. They should not, of course, be run in the same conduit with any other wire or closer than 12 inches to alternating current.

When potentiometer-type instruments are used, thermocouples can be located several hundred feet from the temperature instrument. Millivoltmeter distances cannot be as great. Lead wire of a heavier gage should be used for long distances or two or more lead wires can be connected in parallel.

Testing

The temperature-control instrument can protect furnaces against open thermocouples (or broken thermocouples). Thermocouples should be either changed at intervals determined by experience or tested to protect them against calibration errors.

In testing the thermocouple, the check thermocouple should be inserted in the same tube, or well, or in an adjacent test hole. The check thermocouple should not be used for any other purpose and should be tested frequently against a master standard in a salt pot or small tube furnace. The check thermocouple should be of exactly the same type and size as the one being tested.

Automatic Controls

Pyrometers can be arranged so the moving element of the instrument not only indicates the temperature by its position relative to a scale, but also controls the temperature by regulating the heat supply. Figure 10-20 shows such a control for an electric furnace that uses solid-state devices and is relatively trouble-free. In some cases, even the switching mechanism is a solid-state device. The control can be set for any temperature desired within certain maximum and

Figure 10-20   Automatic control for an electric furnace.
(Courtesy Thermolyn Corp.)

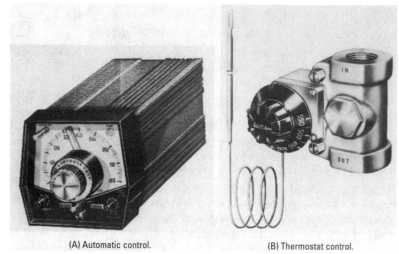

(A) Automatic control.   (B) Thermostat control.

Figure 10-21 Gas furnace temperature controls.
(Courtesy Johnson Gas Appliance Co.)

minimum limits and may be applied to furnaces heated either by gas, oil, or electricity. Figure 10-21A shows such a unit for a gas-heated furnace, while Figure 10-21B shows the thermostat control for a low-temperature gas furnace where the temperature variation can be larger.

Recording Pyrometers

A *recording pyrometer* is provided with some kind of marking device that traces a line upon a chart graduated with reference to time and temperature. Thus, the temperature at any period is shown graphically on the chart. Such a recording pyrometer may be connected to more than one furnace. Different colors can be used to indicate the different furnaces to avoid confusion.

Summary

In general, heat-treating furnaces can be classified with respect to the source of heat (such as gas-fired, oiled-fired, and electric) and to the method of applying heat (such as overfired, under-fired, recuperative, and forced convection). The method of controlling heat (manual or automatic) and the protection of the work (such as a controlled-atmosphere type of furnace) are also important.

Regulation of temperature of the workpiece being treated is the basic factor in modern industrial furnace control. The controls must regulate automatically. This must be done at the lowest possible cost in spite of changes in production, ambient temperature, supply voltage, fuel characteristics, or other changes in the heat-treating process.

Many types of furnaces are designed for a special application or feature (such as high-speed, pusher, rotary-hearth, tilting-oven, pot, box, tool, air-tempering, continuous high-temperature, and carburizing furnaces). Gas-fired furnaces have a substantial frame or casing mounted on cast-iron legs. Heavy steel channels are provided to prevent sagging. Many furnaces have a counterbalanced door and lifting mechanism.

The pot-hardening furnace is designed for indirect heating, the materials being placed in a liquid heat-transmitting medium. Pot furnaces are built for either gas or oil firing. The basic parts of a pot furnace are the furnace, pot, drain, and hood. The burners are arranged to fire tangentially through accurately-formed combustion tunnels into the combustion chamber.

The basic design of an electrically heated pot furnace is to form a single continuous helix of heavy-duty nickel-chromium rod, either round or square in shape (depending on the size of the furnace), around the lining tile of the chamber so that a uniform diffusion of heat is achieved.

Recent developments have been the improvement of the electrode-type pot furnace and the salt baths used in them. Industry is using more of these furnaces because of their many advantages and low maintenance costs. They are especially useful in the heat-treatment of high-speed steel.

A neutral atmosphere is essential for the correct hardening of high-speed steel tools. There, surface changes that can be produced by different furnace atmospheres are scale, decarburization, and carburization. Considerable research has been conducted with a wide range of atmospheres to determine the neutral atmosphere and develop methods of achieving it. Inert gases are being used (especially in electrically heated furnaces) to exclude air from the work being heated.

Much improvement has been made in the controls of heat-treating furnaces. Most such furnaces in industry are being controlled by automatic controls with pyrometers. The desired temperature is set on the control, and the pyrometer holds the furnace at that temperature. A recording pyrometer can also be used if a record is needed of the temperature achieved.

Review Questions

1. How are heat-treating furnaces classified?
2. What are the four types of electrically heated furnaces used for heat-treating?
3. Vertical furnaces of the _____ type are used for heating long, slender work.
4. The _____ furnace is designed for indirect heating, the materials being placed in a liquid heat-transmitting medium.
5. Much of the increased use of the electrode-type furnace is because of the development of _____ baths to use in the furnaces.
6. Cyanide-bearing salt baths are used in heat-treating and case-hardening of _____metals.
7. A _____ atmosphere is essential for the correct hardening of high-speed tools.
8. What is scale?
9. What is meant by decarburization?
10. Most _____controls for heat-treating furnaces are referred to (incorrectly) as pyrometers.
11. What is a pyrometer?
12. Why is absolute temperature difficult to determine?
13. Radiant energy that is given off by a hot surface is a function of the surface temperature and the surface _____.
14. What is the only satisfactory commercial device for furnaces operated at temperatures above 2700°F?
15. What is a thermocouple?
16. How do thermocouples work?
17. Virtually all of the heat received by a thermocouple is by _____.
18. Where can thermocouples be sent for calibration?
19. What does an optical pyrometer do?
20. Why is a recording pyrometer used?

Chapter 11

Annealing, Hardening, and Tempering

In the heat-treatment process of steel, certain changes occur to alter some of its properties. The processes involve heating and cooling the metal, in its solid state, for the purpose of changing its mechanical properties. Steel may be made harder, stronger, tougher, or even softer through various heat-treating processes. The type of steel and its characteristics must be known before it is heat-treated. The material used to make a certain part was selected because of the properties it would possess when the part was finished

The correct temperature is the important factor in any heat-treating process. Improper temperatures in heating do not produce the desired results. Steel will be "burned" at a degree of temperature that is too high. The heat-treating processes are annealing, hardening, and tempering.

The *degree of hardness* of steel can be changed by heat treatment. Hardness is a relative term. All steels have hardness, but some steels are harder than others.

The heat-treating processes consist of heating metal according to a time-temperature cycle, which consists of three steps:

1. Heating the steel to a certain temperature
2. Holding the steel at a certain temperature for a period of time for soaking
3. Cooling at a certain rate

During these three steps, the properties of the steel may be altered in various ways, depending on its chemical content and the process used. These operations usually alter the internal structure in some way.

Annealing

Annealing is a process of softening metal by heating it to a high temperature and then cooling it slowly. Steel is annealed by heating to a low-red heat and letting it cool slowly. The objective of the annealing process is to remove stresses and strains set up in the metal by rolling or hammering so that it will be soft enough for machining. The longer the period that the metal is allowed to cool, the softer the metal will be because more of the stresses and strains will have been removed.

Methods of Annealing

A common method of annealing steel is to place the steel in a cast-iron box, cover it with a material (such as sand, fire clay, ashes, and so on), and then heat in a furnace to a proper temperature. The box and contents are then cooled slowly to prevent any hardening. The sand, clay, and so on, are placed around the steel to exclude air and prevent oxidation, as well as to delay the cooling process.

Annealing with water is not the best of the annealing processes, but it is a quick method. In this process, the metal is heated slowly to a cherry red. Then, it is placed in a dark place and observed until the color is no longer visible; when it reaches this stage, it is cooled in water. The annealed metal is then soft enough to machine. Usually, a piece of steel annealed in this way is much softer than if it is packed in charcoal and cooled overnight.

In machining steel that has been annealed by the common method (that is, charcoal), the chips generally are removed in long close-curled lengths, and the surface presents a torn texture. The chips are torn because the steel is too soft. Water annealing overcomes this defect. The water annealing method is not generally recommended because it is considered to have a deteriorating effect on the steel.

Another method quite often used is to heat the steel to the proper temperature, hold it at this temperature for the right length of time, and then allow it to cool slowly (often in the furnace) until it is under 1000°F (538°C). Then, it is taken out and allowed to cool in air. In general, the higher the carbon content, the slower the cooling rate required.

Temperature for Annealing

The temperature required for annealing of steel is slightly above the *critical point* (the temperature at which internal changes take place in the structure of the metal) of the steel. The critical point varies, of course, with the different steels. Some typical annealing temperatures are:

- Low-carbon steel—1650°F (899°C)
- High-carbon steel—1400°F (760°C) to 1500°F (815.55°C)
- High-speed steel—1400°F (760°C)

Critical Points

There are two critical points in steels of ordinary carbon content. A lower temperature is required to produce the internal changes in structure in a steel containing a higher percentage of carbon than in a steel that contains a lower percentage of carbon. The two critical points are *decalescence* and *recalescence.*

The *point of decalescence* occurs at the temperature where the pearlite changes to austenite as the steel is heated. The *point of recalescence* occurs as the steel is cooled slowly—the austenite returns to pearlite. Thus, the point of decalescence occurs as the temperature is rising, and the point of recalescence occurs as the temperature is falling.

Structure of Carbon Steel

In fully annealed carbon steel, there are two constituents: the element iron in a form known as *ferrite* and the chemical compound iron carbide, known as *cementite*, which has 6.67 percent carbon. Some of the two constituents will be present in a mechanical mixture known as *pearlite*. This mechanical mixture consists of alternate layers of ferrite and cementite. It often has the appearance of mother-of-pearl under the microscope (hence its name). Pearlite contains 0.85 percent carbon, so the amount of pearlite depends on the carbon content of the steel. A low- or medium-carbon steel will consist of pearlite and ferrite. A carbon steel of 0.85 percent carbon is all pearlite. A high-carbon steel over 0.85 percent carbon will consist of pearlite and cementite.

To fully anneal carbon steel, it is heated above the lower critical point (point of decalescence). The alternate bands of ferrite and cementite (which make up the pearlite) begin to merge into each other. This heating process continues until the pearlite is "dissolved," forming what is known as *austenite.* If the temperature of the steel continues to rise, the excess ferrite or cementite will begin to dissolve into the austenite until only austenite will be present. This temperature is called the *upper critical temperature.*

If this steel (all austenite) is now allowed to cool slowly, the process of transformation that took place during the heating will be reversed, but the upper and lower critical temperatures will occur at somewhat lower temperatures. The steel at room temperatures will have the same proportions of ferrite or cementite and pearlite as before. The austenite will have disappeared. Figure 11-1 shows the critical temperatures and the effect of heat on carbon steel.

Steels with less than 0.85 percent carbon must be heated above the upper critical temperature to be annealed or hardened. This is

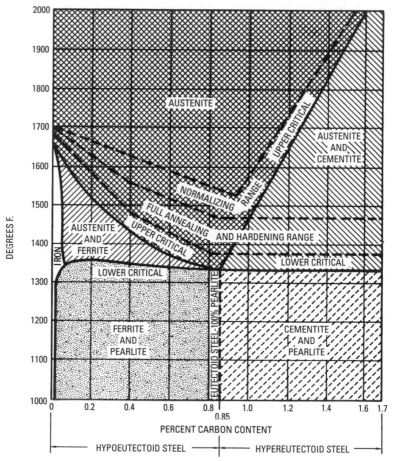

Figure 11-1 Structural constitution of carbon steel. *(Courtesy Machinery's Handbook, the Industrial Press)*

evident when one locates the "full annealing and hardening range" (in Figure 11-1) and compares it to the upper critical temperature.

Effects of Forging

Forging is the process by which steel (or other metal) is *heated and hammered* into the various shapes of tools, machine parts, and so on. In the forging process, the piece that is to be forged should be heated uniformly at the correct temperature. The correct degree of heat depends on the kind of steel and the shape, size, and use that is to be made of the finished product.

The hammering process increases the density of the grains in the steel, which improves the steel. Forging should not take place below a certain temperature because the grains can be crushed, which is damaging to the steel.

If the entire piece of work is not forged, internal strains and stresses are produced in the metal because of the difference in densities of the hammered and unhammered portions of the steel. These strains can be relieved by annealing. If only a portion of a piece is to be forged, the piece should be annealed prior to hardening and tempering. Examples of pieces that are only partially forged are lathe tools, cold chisels, and screwdrivers. These pieces require annealing, hardening, and tempering.

Hardening

The process of *hardening* metal is accomplished by increasing the temperature of a metal to its point of decalescence and then quenching it in a suitable cooling medium. In actual practice, the temperature of the metal that is to be hardened should be increased to slightly above the point of decalescence for two reasons:

- To be certain that the temperature of the metal is above the point of decalescence at all times
- To allow for a slight loss of heat while transferring the metal from the furnace to the quenching bath

Steel with less than 0.85 percent carbon needs to be heated above its upper critical temperature.

Heating Process

Steel can be heated most successfully by placing it in a cold furnace and then bringing the furnace and its charge to the hardening temperature slowly and uniformly. Commercially, only those steels that are difficult to harden are heat-treated in the manner described previously. For example, some tool steels are heated for hardening by first placing them in a preheating furnace at a temperature of 1000°F (537.77°C) or slightly lower. After the steel is heated uniformly to furnace temperature, it is withdrawn and placed in the high-heat furnace at the hardening temperature. When the steel has become heated uniformly at the hardening temperature, it is then quenched to harden it.

During the heating process, the metal absorbs heat, and its temperature rises until the point of decalescence is reached. At this point, additional heat is taken up by the metal. This heat is converted into work to change the pearlite into austenite without an increase in temperature until the process is completed.

This phenomenon can be compared to the *latent heat of steam* in that when the temperature of water is increased to the boiling point, it will absorb an additional quantity of heat without an increase in temperature. The heat is converted into work that is necessary:

- To bring about a change of state from a liquid to a gas.
- To overcome the pressure of the atmosphere in making room for the steam.

The total latent heat required to bring about these changes consists of *internal* latent heat and *external* latent heat.

After the metal has been heated, *quenching* serves to fix permanently the structural change in the metal that causes the metal to remain hard after it has been heated to the point of decalescence. If the metal were not quenched and allowed to cool slowly, the austenite would be reconverted to pearlite as the temperature decreases, causing the metal to lose its hardness. When steel is cooled faster than its critical cooling rate, which is the purpose of quenching, a new structure is formed. The austenite is transformed into *martensite*, which has an angular, needlelike structure and a very high hardness.

Martensite has a lower density than austenite. Therefore, the steel will increase in volume when quenched. Some of the austenite will not be transformed during quenching. This austenite will gradually change into martensite over a period of time. This change is known as "ageing." This "ageing" results in an increased volume, which is objectionable in many items, such as gages. The cold-treating process (discussed in Chapter 15) can be used to eliminate this problem.

The experienced worker can determine the critical point (point of decalescence) on the basis of color. The heated metal should be transferred to a dark place to judge color. Also, metal bars will decrease in length when quenched at a temperature below the critical point in water. If a metal bar is quenched in water at a temperature above the critical point, hardening also will be indicated.

Pyrometers are used in production work to indicate the critical temperature if the critical temperature of a given steel is known.

A magnetic needle can be used to determine whether steel has been heated above the critical point. When heated above the critical point, a piece of steel loses its magnetism. It will attract a magnetic needle if it has been heated to any temperature below the critical point. In making the test with a magnetic needle, the magnetic influence of the cold tongs should be eliminated. A pivoted bar magnet can be introduced into the furnace momentarily to determine the presence or absence of magnetism in the piece of steel.

Heating Baths

Other methods of heating are used in the hardening process. Various liquid baths are used as follows:

- Lead
- Cyanide of potassium
- Barium chloride
- Mixture of barium chloride and potassium chloride or other metallic salts

The chief advantage of the liquid bath is that it helps to prevent overheating of the workpiece. The work cannot be heated to a temperature higher than the temperature of the bath in which it is heated. Other advantages include:

- The temperature can be easily maintained at the desired degree.
- The submerged steel can be heated uniformly.
- The finished surfaces are protected against oxidation.

The *lead* bath cannot be used for high-speed steels because it begins to vaporize at 1190°F (643.33°C). The *cyanide of potassium* bath is used in gun shops extensively to secure ornamental effects and harden certain parts.

A thermoelectric pyrometer can be used to indicate the temperature of a *barium chloride* bath. *Potassium chloride* can be added for heating to various temperatures as follows:

- For temperatures from 1400°F (760°C) to 1650°F (899°C), use three parts barium chloride to one part potassium chloride.
- For higher temperatures, reduce the proportion of potassium chloride.

Quenching or Cooling Baths

Although water is one of the poorest of all liquids for conducting heat, it is the most commonly used liquid for quenching metals in heat-treating operations. Water cools by means of evaporation.

To raise the temperature of 1 pound of water from 32°F (0°C) to 212°F (100°C) requires 180 Btu. To convert it into steam, an additional 950 Btu are required. Thus, the latent heat of vaporization (950 Btu) absorbed from the hot metal is the cooling agency. For efficient cooling, the film of steam must be replaced immediately by another layer of water—this requires circulation. If the heated metal is plunged into the water, thermocirculation results. However, if the metal is placed in the water horizontally, the film of steam that forms on the lower side of the piece is "pocketed," and the cooling action is greatly retarded. In still-bath quenching, a slow up-and-down movement of the work is recommended.

In addition to plain water, several other solutions can be used for quenching. Salt water, water and soap, mercury, carbonate of lime, wax, and tallow can also be used. Salt water will produce a harder "scale." The quenching medium should be selected that will cool rapidly at the higher temperatures and more slowly at the lower temperatures. Oil quenches meet this requirement for many types of steel. There are various kinds of oils employed, depending on the nature of the steel being used.

The quenching bath should be kept at a uniform temperature so that successive pieces quenched will be subjected to the same conditions. The next requirement is to keep the bath agitated. The volume of work, the size of the tank, the method of circulation, and even the method of cooling the bath need to be considered in order to produce uniform results.

High-speed steel is usually quenched in oil. However, air is also used for quenching many high-speed steels. Air under pressure is applied to the work, that is, by air blast.

Tempering

The purpose of *tempering*, or "drawing," is to reduce brittleness and remove internal strains caused by quenching.

The process of tempering metal is accomplished by reheating steel that has been hardened previously and then quenching to toughen the metal and make it less brittle. Unfortunately, the tempering process also softens the metal.

Tempering is a reheating process—the term *hardening* is often used erroneously for the tempering process. In the tempering process, the metal is heated to a much lower temperature than is required for the hardening process. Reheating to a temperature between 300°F (149°C) to 750°F (399°C) causes the martensite to change to *troostite*, a softer but tougher structure. A reheating to 750°F (149°C) to 1290°F (699°C) causes a structure known as *sorbite*, which has less strength than troostite but much greater ductility. Table 11-1 shows tempering temperatures for various tools.

Table 11-1 Typical Tempering Temperatures for Certain Tools

Degrees Fahrenheit	Degrees Celsius	Temper Color	Tools
380	193	Very light yellow	Lathe center cutting tool for lathes and shapers
425	218	Light straw	Milling cutters, drill, and reamers
465	241	Dark straw	Taps, threading dies, punches, hacksaw blades
490	254	Yellowish brown	Hammer faces, shear blades, rivet sets, and wood chisels
525	274	Purple	Center punches and scratch awls
545	285	Violet	Cold chisels, knives, and axes
590	310	Pale blue	Screwdrivers, wrenches, and hammers

Color Indications

Steel that is being heated becomes covered with a thin film of oxidation that grows thicker and changes in color as the temperature rises. This variation in color can be used as an indication of the temperature of the steel and the corresponding temper of the metal.

As the steel is heated, the film of oxides passes from a pale yellow color through brown to blue and purple colors. When the desired color appears, the steel is quenched in cold water or brine. The microscope can be used to explain the phenomena associated with the change in color. Steel consists of various manifestations of the same compound rather than separate compounds.

Although the color scale of temperatures has been used for many years, it gives only rough or approximate indications, which vary for different steels. Table 11-2 shows the color scale for temper colors and corresponding temperatures.

Table 11-2 Temper Colors of Steel

Colors	Temperatures (Fahrenheit)	(Celsius)
Very pale yellow	430	221
Light yellow	440	227
Pale straw-yellow	450	232
Straw-yellow	460	238
Deep straw-yellow	470	243
Dark yellow	480	249
Yellow-brown	490	254
Brown-yellow	500	260
Spotted red-brown	510	266
Brown-purple	520	271
Light purple	530	277
Full purple	540	282
Dark purple	550	288
Full blue	560	293
Dark blue	570	299
Very dark blue	600	316

Specially prepared tempering baths equipped with thermometers provide a more accurate method of tempering metals. These are much more accurate than the color scale for tempering.

Case-Hardening

The case-hardening operation is a localized process in which a hard "skin," or surface, is formed on the metal to a depth of $\frac{1}{16}$ (0.0625) to $\frac{3}{8}$ (0.375) inch. This hard surface, or "case," requires two operations:

- Carburizing where the outer surface is impregnated with sufficient carbon.
- Heat-treating the carburized parts to obtain a hard outer case and, at the same time, give the "core" the required physical properties.

Case-hardening usually refers to both operations.

Carburizing is accomplished by heating the work to a temperature below its melting point in the presence of a material that liberates carbon at the temperature used. The material can be solid (charcoal, coke, and so on), liquid (sodium cyanide, other salt baths), or gas (methane, propane, butane). Often, only part of the work is to be case-hardened. The four distinct ways by which case-hardening can be eliminated from portions of the work are:

- Copper-plating
- Covering the portion that is not hardened with fire clay
- Using a bushing or collar to cover the portion that is to remain soft
- Packing with sand

An article that is to be case-hardened can be copper-plated on the portion that is to remain soft. This is especially useful when a liquid carburizing process is used. The portion to remain soft can also be protected by covering with fire clay, covering with a collar or bushing, or packing the portion in sand. The size and shape of the work will dictate the methods as well as the type of steel and the process to be used.

The case-hardening furnace must provide a uniform heat. Steel that is to be case-hardened must be selected carefully. Since oil and gas have superseded coal as fuels for case-hardening furnaces, furnace construction has changed considerably. Careful consideration should be given the carbonaceous material used in packing the parts and the box in which the material is packed. The operation of packing the insulated workpiece is referred to as *local hardening*.

In the case-hardened process, articles that are to be case-hardened are heated to a cherry red color in a closed vessel along with the carbonaceous material and then quenched suddenly in a cooling bath. Malleable castings can be case-hardened so that they acquire a polish. Malleable iron can be case-hardened by heating it to a red heat, rubbing cyanide of potassium over the surface or immersing it in melted cyanide, reheating, and then quenching it in water. Usually, the hardening operation is a separate operation following the carburizing and is designed for the part and the steel used.

Both iron and steel can be case-hardened, but it is used mostly on steel products. The gears of the transmissions of automobiles are a typical example of case-hardening so that they can withstand the abuse of "shifting gears" (not synchromesh) without waiting for them to synchronize.

Variations on Case-Hardening Methods

A commonly-used method of case-hardening is to first carburize the material and then allow the boxes to cool with the work in them, after which they are reheated and hardened in water. For work such as bolts, nuts, screws, and so on, it is satisfactory to dump them into water directly from the carburizing furnace without reheating.

A common iron wheelbarrow with two pieces of flat iron placed across it lengthwise should be provided. Place a sieve made of ⅛-inch wire, having ¼-inch mesh and approximately 18 inches square by 6 inches in depth, on the bars. The sieve should have a handle ⅝ inch in diameter by 6 feet in length. The boxes are emptied into the sieve, and, after sifting, the heated material is dumped into a tank of cold water, which should be large enough to prevent the water from heating too quickly. Care should be taken not to empty the entire contents of the boxes into the water in one place.

A constant flow of water should be available while the work is being hardened. The work should never be removed from the furnace until the temperature has been lowered. The steel should be treated as tool steel after it is carburized. It is harmful to the steel to harden it at the high carburizing temperature.

Gears and other parts that should be tough, but not extremely hard, should be hardened in an oil bath. The work is less liable to warping, and the hardened product can withstand the shocks and severe stresses without breakage.

Summary

The degree of hardness of steel can be changed by heat treatment. All steels have hardness, but some steels are harder than others. The heat-treating processes of steel are annealing, hardening, and tempering.

The correct temperature is the important factor in any heat-treating process. Improper temperatures in heating do not produce the desired results. Steel will be "burned" at a degree of temperature that is too high.

Annealing is a process of softening metal by heating it to a high temperature and then cooling it very slowly. The objective of the annealing process is to remove stresses and strains set up in the metal by rolling or hammering so that it will be soft enough for machining.

The process of hardening is accomplished by increasing the temperature of a metal to the point of decalescence and then quenching it in a suitable cooling medium. The process of tempering is accomplished by reheating steel that has been hardened previously and then quenched to toughen the metal and make it less brittle. Tempering processes also can soften certain metals.

Review Questions

1. What are the three heat-treating processes?
2. What are the three steps in the heat-treating process?
3. What is annealing?
4. How is annealing done?
5. Why is annealing done?
6. Why isn't water annealing a preferred method?
7. What is the critical point in annealing steel?
8. What is the point of decalescence?
9. What is pearlite?
10. Identify the following terms:
    a. Austenite
    b. Ferrite
    c. Cementite
11. How is forging done?
12. How is metal hardened?
13. What is latent heat of steam?
14. What is quenching?
15. How are pyrometers used in production work?
16. What is the chief advantage of the liquid bath?
17. Why can't the lead bath be used for high-speed steels?
18. What is the purpose of tempering?
19. What is the temperature for tempering screwdrivers?
20. What is the temperature of the following temper colors of steel?
    a. Dark blue
    b. Dark yellow
    c. Brown-purple
21. What does case-hardening do?

22. How is carburizing accomplished?

23. Both iron and _____ can be case-hardened.

24. Gears should be hardened in _____?

25. It is _____ to steel to harden it at the high carburizing temperature.

Chapter 12

Principles of Induction Heating

Induction heating is being used extensively in the hardening of steel. It is particularly applicable to parts that require localized hardening (or controlled depth of hardening) and to irregularly shaped parts that require uniform surface hardening around their contour (such as cams). The following are some advantages of induction heating and hardening:

- Little oxidation or decarburization
- Exact control of depth and area of hardening
- Good regulation of degree of hardness obtained by automatic timing of heating and quenching cycles
- A short heating cycle, which may range from a fraction of a second to several seconds
- Minimum warpage or distortion
- Possibility of using carbon steels instead of more expensive alloy steels

The designer understands the advantage of applying hardening by induction heating to localized zones. Specific areas of a given part can be heat-treated separately. Welded or brazed assemblies can be built up prior to heat-treating if only internal surfaces or projections require hardening. Stresses at any given point can be relieved by local heating. It is important for the toolmaker to understand the principles of induction heating.

A magnetic field is set up around a wire that carries an electrical current (Figure 12-1). If the wire is formed into a coil or a loop, the magnetic field of the coil is intensified (Figure 12-2).

The intensity of the magnetic field is influenced by three factors:

- The amount of current flowing through the wire
- The number of turns or loops in the coil
- The medium within and surrounding the coil (Figure 12-3)

Since the number of turns in the coil influences the intensity of the magnetic field, it is logical that a coil with several turns carrying a small current can produce the same effect as a coil with a single

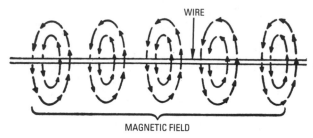

WIRE

MAGNETIC FIELD

Figure 12-1 Diagram showing the magnetic field created around a wire carrying current.

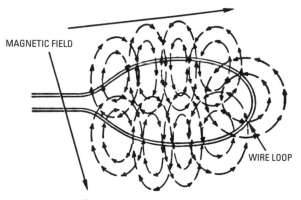

MAGNETIC FIELD

WIRE LOOP

Figure 12-2 Diagram showing the increase in intensity of the magnetic field when a coil or loop is introduced in the wire carrying a current.

turn carrying a large current. Likewise, if the medium is a poor conductor of magnetic lines of force (air, for example), the magnetic field is weak. The magnetic field is intensified if the medium is a good conductor of magnetic lines of force. When a soft-iron bar is placed in a coil that is carrying a current, an electromagnet is produced (see Figure 12-3).

An alternating current flowing through the coil of wire causes the magnetic lines of force to expand and contract. If the magnetic lines of force cut an object that is a conductor of electricity, voltage is induced in the object as the lines of force expand and contract. The transformer, which is used for the purpose of either increasing or decreasing voltage, is an example of a practical application

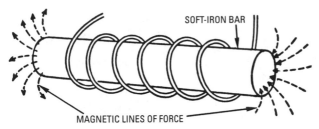

Figure 12-3   Diagram showing the increase in intensity of the magnetic field when the wire carrying the current is wound around a soft-iron rod.

of the preceding phenomenon (Figure 12-4). In the transformer, a secondary coil having a fixed number of turns of wire is placed in the magnetic field of a primary coil having a different number of turns of wire. When a low-frequency alternating current is passed through the primary coil of the transformer, an alternating current of different voltage is induced in the secondary coil. The characteristics of the induced current are determined by the number of turns of wire in the two coils.

Induction heating is a result of application of the aforementioned transformer principle. Since the object to be heated is made of metal and is a conductor of electricity, electrical current can be induced within the object in the same manner that it is induced in the secondary coil of a transformer. The resistance of the metal to the flow of these induced currents results in rapid heating action that is

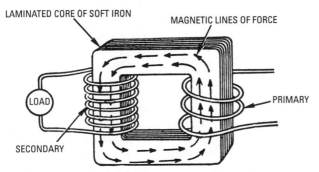

Figure 12-4   Note the effect of the laminated soft-iron core on the magnetic lines of force indicated by arrowheads. This is the basic principle of a transformer.

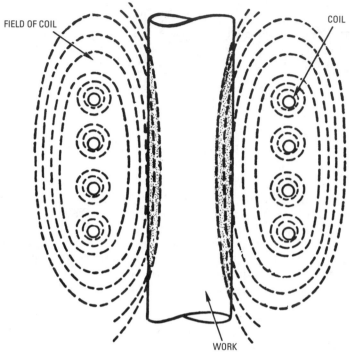

Figure 12-5 Diagram showing the electromagnetic field created by a load coil in induction heating.

often called *eddy current losses* (Figure 12-5). If the object is made of steel or iron, it is magnetized, demagnetized immediately, and remagnetized in the opposite direction, as the current alternates in direction.

At the high frequencies used in induction heating, molecular friction is created by the rearrangement of the steel or iron molecules, as the magnetic state is changed rapidly and continuously. This molecular friction generates heat within the object. The phenomenon is known as heating by *hysteresis* losses. Heat energy can be induced in a piece of steel at the rate of 100 to 250 Btu per square inch per minute by induction heating. This compares with a rate of 3 Btu per square inch per minute for the same material at room temperature when placed in a furnace with a wall temperature of 2000°F (1093.33°C). This means a very short heating cycle is needed.

The amount of current (amperes) changes constantly as ac flows in a wire or conductor. The amount of current fluctuates continually from zero to a positive maximum and back to zero again. Then, the amount of current changes to negative maximum and returns to zero. This complete reversal of current is called a *cycle*. If this cycle occurs in 1 second, it is called a *hertz*. The number of cycles that occur in 1 second is referred to as the *frequency* of the current. The frequency of current is usually 60 hertz (cycles per second).

Usually, the higher the frequency used, the shallower the depth of heat penetration. Low power concentrations and low frequencies are usually employed for heating clear through, for deep hardening, and for large workpieces. Currents at high frequencies are used for shallow and controlled depths of heating (such as surface hardening, and in localized heat treating of small workpieces).

The depth of heat penetration and the frequency used can be illustrated. A ½-inch round bar of hardenable steel can be heated through its entire structure by an induced current of 2000 hertz so it can be hardened, but a ¼-inch bar of the same steel would not reach a sufficiently high temperature to permit hardening. At 9600 hertz, the ½-inch bar can be surface-hardened to a depth of 0.100 inch, while the ¼-inch bar can be through-hardened. The ¼-inch bar would require a current at over 100,000 hertz for surface hardening.

Hardening, annealing, and normalizing by induction, low power concentrations, and low frequencies are desirable. This will prevent too great a temperature difference between the surface and the interior of the work. Widening the spacing between work and applicator coil is one way of reducing the amount of power delivered to the work.

Figure 12-6 shows various shapes of heating coils that are used for high-frequency induction heating. Alternating current has a tendency to concentrate on the surface of a conductor. This tendency is not noticeable with 50- or 60-hertz current, but it is very pronounced with higher frequencies. Since the center of the conductor carries a very small amount of current, the load coil can be made of hollow copper tubing in which water can be circulated to carry away heat that is generated by the flow of the current and radiant heat from the heated object.

Figure 12-7 shows an automatic rod-heating device prior to forming.

Figure 12-6   Various types of coil shapes used for high-frequency induction heating.

Figure 12-7   An automatic rod-heating device, using radio-frequency induction heating. *(Courtesy Lepel Corporation)*

Adjustable Induction Heating Coil

The Lewis Research Center in Cleveland, Ohio, in conjunction with NASA, has designed an adjustable induction heating coil using three coils that can be adjusted independently to obtain the desired distribution of temperature.

Adjustable
positioning
Blocks

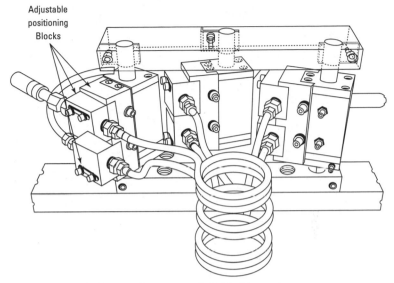

Figure 12-8   Induction heating coil divided into three segments.
(Courtesy NASA)

The induction heating method has been used for more than 30 years, but the control of the heat over a piece of material inserted into the coil has been less than desired. The newer approach uses a coil in three segments (Figure 12-8). Each segment is attached to a positioning mechanism that enables independent adjustment along both the vertical (axial) and radial (in this case, horizontal) axes. Short lengths of copper tubing at the rear of the assembly are used to connect the three segments of the coil in series. The assembly is connected to the induction heater power supply by two flexible power leads.

Both a 5-kilowatt radio-frequency (450 kilohertz) and a 50-watt audio-frequency (9.6 kilohertz) induction heat have been used successfully with this fixture for many high-temperature material deformation tests. Temperature profiles in these experiments vary only ±3°C from the nominal test temperatures over a one-half inch (13 mm) specimen gage length.

It is worth noting that the concept of the segmented coil allows up to three different coil diameters to be used without difficulty to form a work coil. Also, the direction of the coil winding can be reversed with a minimum of trouble (Figure 12-9).

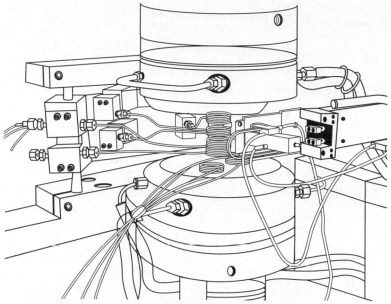

Figure 12-9 Induction heating coil mounted in a 5 kW three-segment coil unit. *(Courtesy NASA)*

Summary

Induction heating has come into use because of the need to have local hardening or controlled depth of hardening. There are several advantages of induction heating and hardening.

The intensity of a magnetic field is influenced by three factors: the amount of current flowing, the number of turns of wire in a coil, and the atmosphere within and surrounding the coil.

Since the number of turns in the coil influences the intensity of the magnetic field, it is logical that a coil with several turns carrying a small current can produce the same effect as a coil with a single turn carrying a larger current. The magnetic field is intensified if the medium through which it passes is a good conductor of magnetic lines of force. When a soft-iron bar is placed in a coil that is carrying current, an electromagnet is produced.

If the current is alternating, the electromagnet becomes a transformer. Induction heating is an application of the transformer principle. In general, the higher the frequency, the shallower the depth of heat penetration is. High-frequency induction heating is a good way to surface-harden a piece of hardenable steel.

A newer approach to induction heating has been designed with three coils and two frequencies used for the power source. This way, the heating effect can be controlled. It is worth noting that the concept of the segmented coil allows up to three different coil diameters to be used without difficulty to form a work coil.

Review Questions

1. Where is induction heating used?
2. What are the six advantages of induction heating?
3. What three factors affect the intensity of the magnetic field in the induction coil?
4. What is the type of current used in the coil?
5. What is meant by the transformer principle?
6. What is hysteresis?
7. How hot can metal be heated in an induction coil?
8. Widening the spacing between the work and the applicator coil is one way of reducing the amount of power delivered to the _____.
9. Why can the load coil be made of copper tubing?
10. How long has the induction heating method been used?

Chapter 13

High-Frequency Induction Heating

Operations such as annealing, hardening, brazing, soldering, and melting can be performed efficiently with induction heating equipment. Certain characteristics of alternating current (ac) must be considered for an understanding of high-frequency heating.

As you will recall from the discussion in Chapter 12, the amount of current (amperes) changes constantly as ac flows in a wire or conductor. The amount of current fluctuates continually from zero to a positive maximum and back to zero again; then, the amount of current changes to negative maximum and returns to zero. This complete reversal of current is called a *cycle*. If this cycle occurs in 1 second, it is called a *hertz*. The number of cycles that occur in 1 second is referred to as the *frequency* of the current. The frequency of current is usually 60 hertz (cycles per second).

High-frequency heating simply indicates that the number of hertz (frequency) of the current supplied normally from the power lines of the utility company has been greatly increased by means of a generator, spark gap, or vacuum tube equipment (Figure 13-1).

Producing Heat by Resistance

Almost all objects that are electrical conductors can be heated if they are exposed to a strong electromagnetic field. This field is produced by a strong electric current flowing through the load coil.

In general, metals are good conductors of electricity, and their surface resistance is of such a nature that an appreciably high current (*eddy current*) can flow at the surface of the workpiece. These eddy currents are dependent on the frequency of the equipment, the surface resistance of the metal, and the power applied. If any of these factors is increased, a greater quantity of heat is produced in the metal. Low frequencies require a close coupling between the coil and the work for transferring energy. High frequencies permit a wider spacing between the coil and the load.

Most of the standard types of steel that can be hardened can be hardened by induction heating. Low-carbon steels with a carburized case, as well as medium- and high-carbon steels, may be used in this process.

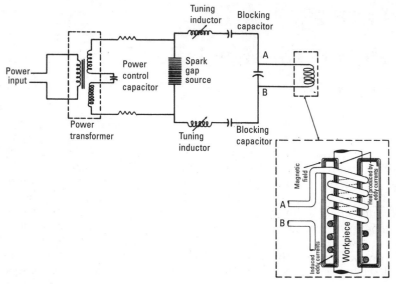

Figure 13-1   Circuit for high-frequency heat-treating furnace.
(Courtesy Lepel Corporation)

Heating Units

High-frequency current advanced from the hypothetical stage to practical realization as early as 1842, when Professor Henry discovered that the discharge of an electrical condenser across a gap was oscillatory. Years later, as Marconi was making use of this important discovery in developing wireless telegraphy, Egbert von Lepel, a noted scientist who recognized the shortcomings of spark gaps used by Henry and Marconi, invented what he called the *quenched spark gap*.

Frequencies of 1000 hertz to 15,000,000 hertz are used in high-frequency induction heating units. Alternating current, because of its inductive effect, tends to concentrate on the surface of a conductor. This tendency is low with 60-hertz current, but it becomes very pronounced at the higher frequencies of induction heating.

Induction heating processes are divided into low-frequency and high-frequency types. For several years, 60-hertz current and currents with frequencies up to a few thousand hertz were used in furnaces to melt metal. Newer installations in industry are mostly of the high-frequency type (Figure 13-2).

Figure 13-2   Automatic vertical scanner for hardening rods, shafts, axles, and spindles. *(Courtesy Lepel Corporation)*

There are three types of equipment used by industry to produce high-frequency current for induction heating:

- Motor generator sets, which provide current at frequencies between 1000 and 10,000 hertz
- Spark gap oscillator units for frequencies between 80,000 hertz and 300,000 hertz
- Vacuum-tube oscillator sets for frequencies ranging from 350,000 hertz to 15,000,000 hertz or more (Figure 13-3)

High-Frequency Applications

High-frequency induction heating reduces the time required for annealing, hardening, stress relieving, brazing, soldering, and forging of metals from several hours to a few minutes or seconds (see

Figure 13-3   A 45-position rotating table for tempering hand tools.
(Courtesy Lepel Corporation)

Figure 13-3). In the process of hardening metals, heat can be applied where it is needed, at the temperature required, and for the required length of time. The surfaces of metal parts, such as crankshafts, rolls, pins, gears, valve stems, valve seats, and the internal surfaces of engine cylinders, can be hardened without affecting the toughness of the metal beneath the surface (see Figure 13-4). High-frequency brazing gives a smooth, tight, and strong union of the parts. In forging, the chief advantage of high-frequency heating is in the short heating cycle.

Quenched spark-gap induction heat-treating units are made in sizes ranging from 4 to 30 kilowatts input and larger. The larger units are provided with a device that corrects the power factor to almost unity.

High-frequency energy is generated in the unit and applied to the object being heat-treated through *work coils*. The work coils are made of copper tubing formed into a coil having either a few turns or many turns. The work coils are connected to the outside of the

Figure 13-4   A two-position station and solid-state generator for hardening hand tools. *(Courtesy Lepel Corporation)*

unit, and they are cooled with water flowing through the copper tubing.

In the heat-treating process, the objects or machine parts that are being heat-treated are placed within the turns of the work coil. Thus, they are exposed to the field of the high-frequency current, producing the heat energy required to heat the object or a particular section where heating is required. Heat energy is induced only in the area of the object that lies within the turns of the work coil. Since the heating action occurs with extreme rapidity (almost instantaneously), it is relatively simple to apply the work coils to a single section of a machine part, or simultaneously to two or more sections of the part, and to heat those sections in such a manner that distortions or structural changes do not occur in any other section of the heated part.

High-frequency current concentrates the heat that it induces on the outer surface of the machine part being heated; the higher the frequency, the more pronounced the "skin effect." Another advantage that can be obtained with high-frequency current is that the work coils are not required to be fitted closely around the object being treated. This is an advantage because the same coil can be used to heat several different objects of different shapes and sizes. Therefore, oddly shaped coils are not necessary for heating irregularly shaped objects.

The smaller the work, the thinner the section (or the shallower the depth to be hardened), and the frequency required will be higher. A short heating period to prevent overheating adjacent areas may require a high power concentration.

Induction heating of internal surfaces can be done by using coils shaped to match the cross section of the opening. For long holes (or if the power available is not sufficient), a short coil (often only one turn) is used, and either the coil or the work is moved so that the heated zone passes progressively from one end of the hole to the other. For a small hole, a hairpin-shaped coil is often used, and the work is rotated to ensure even heating.

After induction heating, the work needs to be quenched in order to be hardened. Quenching may be done by immersion (usually in oil), by a liquid spray (usually water), or by self-quenching. Self-quenching is limited to special cases where the section heated is small or thin, and the mass of the rest of the piece is great enough to cool the heated section. Quenching by immersion offers the advantage of even cooling and is necessary for through-heated parts. Spray quenching is most satisfactory for surface hardening. The quenching unit is often adjacent to the coil, and the quenching cycle can follow the heating cycle without removing the work from the holding fixture. Automatic timing is often applied to both the heating and quenching cycles. This allows the exact degree of hardness to be secured.

Before operating the high-frequency unit, first determine the required temperature for the particular heat-treating operation involved and the time cycle required to produce the correct temperature. If manual control of the unit is to be used, the procedure is as follows:

1. Place the metal part inside the work coil.
2. Step on the floor control pedal to switch on the current and apply heat to the metal part almost instantly.
3. Maintain current for the required number of seconds.
4. Switch off the current.
5. Remove the part.
6. Quench the part to complete the operation.

An automatic timing device can be used to take the place of manual control by the operator. Water, brine, oil, or other media can be used as a quenching medium.

Most of the advantages of induction heating and hardening have been claimed for the induction hardening of gear teeth. Spur gears are the easiest to harden by induction heating. The gear is usually placed inside a circular coil, which is combined with a quenching ring. The process is controlled by an automatic timer. During the heating cycle, the gear is usually rotated at 25 to 35 rpm to ensure uniform heating. With bevel gears, the coil must conform to the face angle of the gear. Some spiral bevel gears tend to heat one side of the tooth more than the other. Often this can be overcome by applying slightly more heat to ensure the hardening of both sides.

Where the workpiece is too long to permit heating in a fixed position, progressive heating may be used. A rod or tube of steel may be fed through a heating coil so that the heat zone travels progressively along the entire length of the workpiece. The quenching ring can be placed next to the coil so that the entire operation is automatic. Gear rack teeth are often hardened in this fashion.

Summary

Operations such as annealing, hardening, brazing, soldering, and melting can be performed efficiently with induction heating equipment. Almost all objects that are electrical conductors can be heated if they are exposed to a strong electromagnetic field.

In general, metals are good conductors of electricity, and their surface resistance is of such a nature that an appreciably high current can flow at the surface of the workpiece. Low frequencies require a close coupling between the coil and the work for transferring energy. High frequencies permit a wider spacing between the coil and the load or work.

Higher frequencies are needed for shallow depth of hardening, for smaller work, and for thin sections. The heating coils can be shaped to fit the work being heated. After induction heating, the work must be quenched in order to be hardened. Quenching may be by immersion, by spray, or by self-quenching. Self-quenching is limited to special cases where the section heated is small or thin and the mass of the rest of the piece is great enough to cool the heated section. Automatic timing can be applied to both heating and quenching.

Many gears are induction-hardened. Spur gears with automatic timing cycles present few problems. Progressive heating can be applied to long workpieces.

Review Questions

1. What operations can be done with induction heating?
2. What is frequency?
3. What is hertz?
4. What objects can be heated by induction heating?
5. What are eddy currents?
6. What did Egbert von Lepel discover?
7. What are the three types of equipment used today to produce high-frequency current for induction heating?
8. Why is high frequency preferred in brazing?
9. What is the chief advantage of high-frequency heating in forging?
10. Where are the objects being heat-treated placed when utilizing the induction heating process?
11. What is high-frequency induction?
12. What operations can be performed by high-frequency induction?
13. What is a quenched spark gap?
14. Explain why a high frequency is needed for surface hardening.
15. What is progressive heating?
16. How is quenching done?
17. Why is a spur gear usually rotated while being heated?

Chapter 14

Furnace Brazing

A number of well-established methods of fabricating and forming metal parts of assemblies are used in the manufacturing industry. Each method has certain manufacturing advantages that cannot be provided by other methods. The method usually employed is the one that provides the lowest overall cost for making and servicing a particular product, considering all factors.

Electric-furnace brazing has become well established among the various methods of forming parts and fabricating assemblies. This method can provide certain benefits that can be obtained by no other method. It has improved quality in most instances where the method has been adopted. The electric-furnace brazing process usually has been adopted to:

- Replace another method.
- Augment a former production method.
- Produce a new product that has been developed which would be either difficult or impossible to produce by any method other than the furnace brazing process.

Manufacturers of automobile and refrigerator parts make up a large portion of the users of electric-furnace brazing. Great strength in the joints is important to them, and they are also attracted by the tightness, uniformity, and excellent appearance of furnace-brazed subassemblies. Practically all automobiles and many brands of refrigerators contain several furnace-brazed parts. This is also true for many adding machines, accounting machines, cash registers, typewriters, radio receiving sets, and sewing machines.

Basic Process

Brazing is a metal-joining process that uses a nonferrous filler with a melting point above 800°F (427°C) but below that of the base metals. The filler metal wets the base metal when molten and flows between the close-fitting metals because of capillary attraction.

Assemblies are put together with brazing metal, usually in the form of wire, applied near the joints that are to be brazed (Figure 14-1). Then, the assemblies are passed through an electric furnace in which a reducing atmosphere prevents the metals from oxidizing,

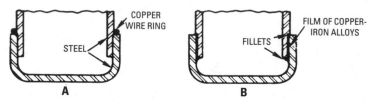

Figure 14-1 Diagrams showing a parts assembly (A) before furnace brazing and (B) after furnace brazing.

frees the metals from any oxides that are present, and thus prepares the surfaces of the parts to be wetted by the molten brazing metal. When the brazing metal melts, it creeps on the surfaces of the parts and is drawn into the joints by capillary attraction to form alloys with the metals in the workpiece. The alloys solidify and develop great strength on transfer of the workpiece to a controlled-atmosphere cooling chamber. The assemblies cool to a temperature at which it is safe for them to contact the outside air without danger of discoloration because of oxidation. The assemblies are delivered from the furnace with strong, tight joints and with clean, bright surfaces.

Electric-furnace brazing has been substituted for torch brazing, dip brazing, soft soldering or sweating, pinning, riveting, welding, machining from solid stock, casting, and forging. Generally, electric-furnace brazing has been employed to replace or augment one of the preceding methods because of certain objections. Following are some objections in these methods that can be overcome by electric-furnace brazing:

- Low strength of parts (they work loose in service).
- Nonuniform strength, which results in uncertain length of life of parts.
- Surfaces become oxidized or covered with flux and require constant cleaning.
- Distortions from localized heating, which require subsequent straightening or machining.
- High cost of forming, which requires machining, patterns, molds, dies, and so on.
- Low production rate, which causes slow manual operations.

Another type of furnace that can be used for brazing is the salt-bath pot furnace. The heating is provided by dipping the parts into a bath of molten salt, which is heated above the melting temperature of the brazing metal.

Salt-bath brazing has four major advantages:

- The work heats rapidly as it is submerged and in complete contact with the salt bath.
- The salt bath protects the work from oxidation.
- Thin pieces can easily be attached to thick pieces without danger of overheating.
- The process can be easily adapted to a conveyor system of production work. (Chapter 10 discusses pot furnaces and salt baths.)

Holding Assemblies Together

The relation of parts in assemblies that are to be brazed must be maintained from start to finish as the assemblies pass through the furnace. There are a number of ways of doing this, and a combination of several different methods may be necessary if the assembly is complicated. The success of a furnace-brazing job depends largely on the methods of holding the assemblies together, the design of the joints, and the means of applying the brazing metal.

The force of gravity is an important factor in furnace brazing. Assemblies of parts tend to fall apart when they are heated because expansion loosens the joints. The brazing metal naturally tends to flow downward more than in other directions. Brazing metals will creep horizontally or in an upward direction on the surfaces of the metal. If applied in excess quantities, they will flow downward quite freely and collect at the low spots. In designing electric-furnace brazing assemblies, it is important to keep in mind the method of holding the assembly together inside the furnace and of setting it up inside the furnace to direct the flow of the brazing metal into the joints to give minimum distortion or movement of parts. Generally, cut-and-trial methods can be used to determine the proper procedure so that the brazing metal can be made to flow into all joints, leaving neat fillets, clean surrounding surfaces, and a job without distortion of parts.

Copper is the most common brazing metal, and steel is the most common parent metal. Other brazing metals and parent metals are also used, but the combination of copper and steel is the most common one.

Laying and Pressing Parts Together

The simplest method of joining two parts is to lay one part on top of the other with brazing metal either placed between them or wrapped around one of the members near the joint. In this method, the weight of the upper member ensures good metal-to-metal contact. The chief

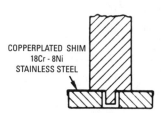

COPPERPLATED SHIM
18Cr - 8Ni
STAINLESS STEEL

Figure 14-2 A shoulder machined on one member is used to make a joint stable for the electric-furnace brazing process.

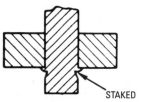

STAKED

Figure 14-3 Staking method used to hold parts assembly in position while it is being brazed in the electric furnace.

disadvantage of this method is the lack of a means of indexing or keeping the parts from moving in relation to one another.

The most common method of assembling parts for electric-furnace brazing is to use a sleeve fit to keep them together. Regardless of the tightness of the sleeve fit, a means must be found to prevent slippage of the parts when they become heated in the furnace.

A shoulder can be machined on one member (particularly if the joint has a vertical axis) to prevent slippage and accomplish stability (Figure 14-2). A punch can be used to turn up burrs in the shaft, as shown in an exaggerated manner in Figure 14-3. This is called *staking*, and it can be used to lock two members together effectively. The staking method is used to retain the indexing of cam, lever, and gear assemblies on shafts or hubs. This can also be accomplished by tack welding (Figure 14-4) and by pinning.

Staking can be omitted where the point that is to be brazed has a horizontal axis and one part does not have a tendency to slide on the other. A snug

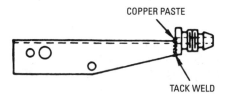

COPPER PASTE

TACK WELD

Figure 14-4 Tack welding can be used to hold the tip to the shank of the electrode holder for brazing in the electric furnace.

fit is necessary. The brazing metal can be placed on the side that is more likely to draw the molten brazing metal into the joint by capillary attraction before it can flow away (Figure 14-5). An added advantage in flow of metal can be gained by tipping the horizontal axis slightly.

When brazing steel with copper, an effort should be made to have a snug fit if possible. The usual tolerances for machined parts cannot be avoided in pressing parts together.

An inexpensive snug fit can sometimes be achieved by straight-knurling the male member (Figure 14-6). The outside diameter is considerably oversize with the hole, but wide tolerances can be used with this arrangement on both the shank and the hole, contributing toward a lower-cost job.

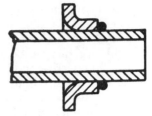

Figure 14-5   A simple press fit can sometimes be used. The brazing metal is placed on the side that is most likely to draw the molten brazing metal into the joint by means of capillary attraction.

Spot welding (Figure 14-7) is another method used frequently to hold assembled parts for electric-furnace brazing. This is a fast and inexpensive method, and a neat job is the usual result. Spot welding

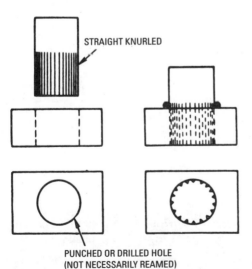

STRAIGHT KNURLED

PUNCHED OR DRILLED HOLE
(NOT NECESSARILY REAMED)

Figure 14-6   Straight knurling on the male member can sometimes give a snug fit to hold the parts together for electric-furnace brazing.

prevents slippage of parts, and dimensions can be held accurately.

Parts can be swaged to hold them together for furnace brazing (Figure 14-8). This is an effective and inexpensive method of assembling spuds into holes in shallow bodies. A strong, tight, and dependable bond that is leakproof and can withstand high pressures is obtained.

Spinning or pressing the end of a tenon into a countersunk hole can be used to lock the parts together with little or no effect on the size of the hole (Figure 14-9). The same result can be obtained by flaring the tenon in a press. The hole is chamfered so that the end of the tenon can be spun or pressed into the chamber. This is a commonly used method for assembling parts that were formerly drilled or pinned.

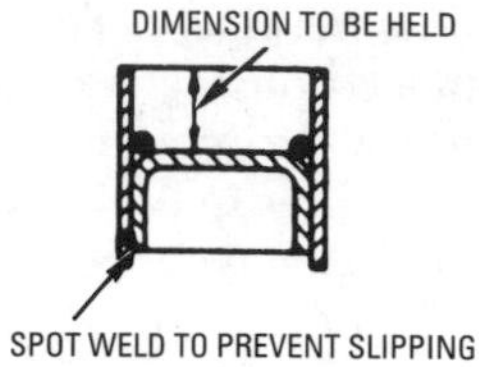

Figure 14-7 Spot welding is used frequently to hold assembled parts for electric-furnace brazing.

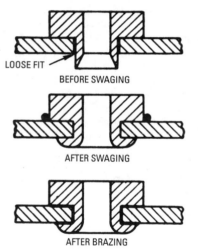

Figure 14-8 Swaging spuds into shells is an effective and inexpensive means of holding spuds in holes in shallow bodies.

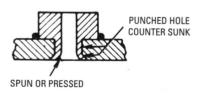

Figure 14-9 A method of holding parts together by spinning.

When tubular members are pressed into headers, an expanded operation can be used to lock the assemblies together for furnace brazing (Figure 14-10). Copper rings can be placed over the tubes after the expanding operation, against the outside of the tube sheet. Alternatively, they can be placed over the tubes before assembly and moved against the inner surface of the tube sheet after the expanding operation, assuming that the rings are accessible.

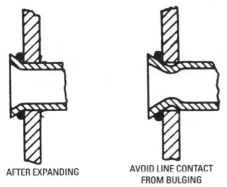

AFTER EXPANDING   AVOID LINE CONTACT
FROM BULGING

Figure 14-10 Tubular members can be expanded into holes or headers to lock the assemblies together for furnace brazing.

Care must be taken in the expanding operation to prevent bulging of the tube to produce only a line contact (see Figure 14-10). A leader on the expanding tool can be used to project into the tube to support the tube wall while the end of the tube is being flared outward.

One or more members of an assembly sometimes can be peened to hold the members together for brazing in the furnace (Figure 14-11). Two stampings are pressed together and then peened around the edges with an air hammer. Copper in the molten state can be sprayed on the parts by means of an oxyacetylene spray gun so that it will be within the joint after assembly.

A disc in the end of a shell can be held in place by crimping the ends of the shell, as illustrated in Figure 14-12. Indentations around the shell can be used as stoppers against which the disc can be held. It is better practice to set such an assembly on end so that the brazing metal can flow downward through the joints. This is often impractical because of lengths. If the assembly must be laid on its side, an oversized ring of brazing metal can be sprung in place as near the joint as possible. A copper-powder paste can be daubed on

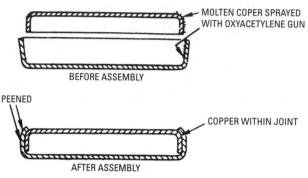

Figure 14-11   The edges of stampings can be peened to hold snug joints as the assembly is brazed in the furnace.

the wire and the adjoining steel surfaces. The hardened paste tends to hold the copper wire in place so that it will not sag away from the joint at the top when it gets hot. The copper paste serves as an auxiliary supply of brazing metal and ensures the presence of brazing metal where it is needed in the joint.

Parts assemblies can also be riveted together. A fan-wheel assembly is a typical example of parts that were formerly riveted but are now riveted and then brazed in the electric furnace.

Parts can be held in their indexed positions for brazing by pinning (Figure 14-13). Tapered pins are driven through small levers pressed on a shaft to prevent movement during the brazing operation.

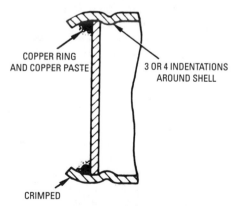

Figure 14-12   Crimping the ends of a shell to hold a disc and the copper ring in position for furnace brazing.

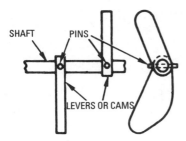

SHAFT

PINS

LEVERS OR CAMS

Figure 14-13 Tapered pins can be used to hold levers or cams in their proper indexed positions for furnace brazing.

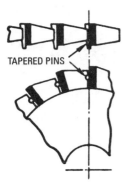

TAPERED PINS

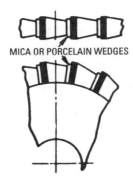

MICA OR PORCELAIN WEDGES

Figure 14-14 Tapered pins (left) and mica or porcelain wedges (right) can be used to wedge tungsten carbide bits into the slots of milling cutters for furnace brazing.

Wedges are also used to lock parts or members into the assemblies (Figure 14-14). Two methods of wedging are used to wedge tungsten carbide bits into the slots of milling cutters for brazing in the electric furnace. Tapered pins and wedges of mica or porcelain can be used to wedge the bits into the slots.

Screws are commonly used to hold parts together for brazing in the electric furnace (Figure 14-15). The manner of putting the parts together is shown in the illustration.

Several methods of overlapping members are effective in furnace brazing (Figure 14-16). Copper clips can be placed at intervals over the upper edge. Thus, brazing metal can be supplied to a joint that is inaccessible.

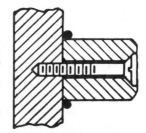

Figure 14-15 Using screws to hold parts together for furnace brazing.

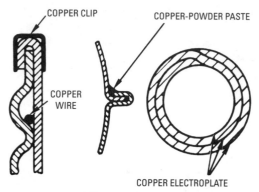

Figure 14-16 Types of overlapping joints suitable for furnace brazing.

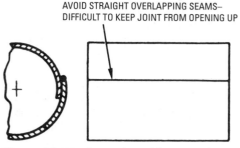

Figure 14-17 An undesirable type of overlapping joint that should be avoided in furnace brazing because of the difficulty in keeping it tight.

A similar method of overlapping is sometimes used on hollow containers (*B* in Figure 14-16). The copper can be placed inside the joint, or copper-powder paste can be applied along the outside of the seams, as indicated in Figure 14-16.

A cross section of copper-brazed, double-walled steel tubing that is made by laterally rolling a copper-plated steel strip is shown in Figure 14-16. This type of tubing is used in gasoline lines of automobiles, oil lines, and hydraulic brake lines.

Figure 14-17 shows an undesirable type of overlapping joint that should be avoided. In long straight seams, it is almost impossible to retain surface contact within the joints because of warpage, resulting in opened or bulged joints that will leak.

Interlocked joints are used frequently in forming assemblies for furnace brazing (Figure 14-18). The edges of steel tubing can be rolled and interlocked to prevent their opening up while passing through the

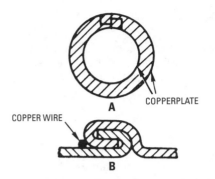

Figure 14-18 Types of interlocking joints that are suitable for electric-furnace brazing.

COPPERPLATE

A

COPPER WIRE

B

furnace. A thin copper plating or wire can be used to supply brazing metal (see Figure 14-18A). If enough time is allowed, the copper brazing metal can creep up the inner sides of the tube and braze the joint even though the seam is located at the top.

An interlocked joint that is sometimes used in making hollow tubing or hollow containers is shown in Figure 14-18B. This type of

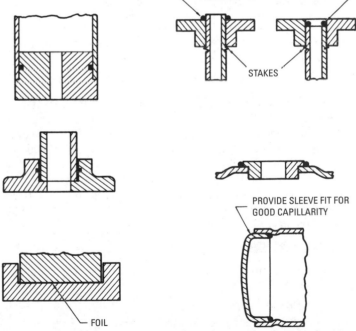

PROVIDE SHOULDERS IF POSSIBLE

STAKES

PROVIDE SLEEVE FIT FOR GOOD CAPILLARITY

FOIL

Figure 14-19 Other examples of good joints for furnace brazing.

joint provides a secure locking effect with little possibility of its opening up when it is heated.

Figure 14-19 shows some other examples of good joints for furnace brazing. Many variations of the examples given in this chapter will be found in industry, as well as other joints that require a fixture or some other method of holding the parts in place during the brazing.

Summary

Electric-furnace brazing has become well established among the various methods of forming parts and fabricating assemblies. It has improved quality in most instances where the method has been adopted. Great strength in joints has been developed, plus tightness, uniformity, and excellent appearance of furnace-brazing sub-assemblies. Practically all automobiles and many brands of refrigerators contain several furnace-brazed parts. The salt-bath pot furnace is also being used for furnace brazing.

Furnace brazing has been used as a substitute for torch brazing, dip brazing, soft soldering or sweating, pinning, riveting, welding, machining from solid stock, casting, and forging. The success of a furnace-brazing job depends largely on the method of holding the assemblies together, the design of the joints, and the means of applying the brazing metal. Copper is the most common brazing metal, and steel is the most common parent metal. Other brazing metals and parent metals are also used, but the combination of copper and steel is the most common one used.

The simplest method of joining two parts is to lay one part on top of the other with brazing metal either placed between them or wrapped around one of the members near the joint. In this manner, the weight of the upper member ensures good metal-to-metal contact. Many problems develop in the assembling of parts. One common method of assembling parts for furnace brazing is to use a sleeve fit to keep parts together. Regardless of the tightness of the sleeve fit, a means must be found to prevent slippage of parts when they become heated in the furnace.

Review Questions

1. What are the three reasons why electric-furnace brazing is being used?
2. What is brazing?
3. Which industries use electric-furnace brazing?

4. Why has electric-furnace brazing been used to replace or augment other methods?

5. Another type of furnace that can be used for brazing is the _____ pot furnace.

6. What is the most common brazing metal?

7. What is the chief disadvantage of laying and pressing parts together for brazing?

8. Why is staking used?

9. Why is spot welding used during the brazing process?

10. Interlocked joints are used frequently in forming _____ for furnace brazing.

Chapter 15

Cold-Treating Process

Cold treating is a comparatively new process with many possible applications. Chilling machines are built for high-production work, and they are standard equipment in plants that use the cold-treating process.

Cold treating is used as a supplement to heat treating in the process of hardening metals. A general knowledge of both the structural changes and the essential processes of heat treatment is required to understand fully the effect of subzero temperatures on tool steels.

Fundamental Principle of Cold Treating

In accordance with the slip-interference theory, *hardness* is described as a resistance to permanent deformation of the material. When either tensile or compression forces are applied to a piece of metal, the result is a slipping action along the boundaries of the grain or crystallographic planes.

The amount of slip or slide that actually occurs is dependent on the grain size, slip-interference particles, and the force brought to bear. Decreasing the size of the particles increases the number of interference particles and increases the resistance to slip, resulting in more balance.

Decalescence

As steel is heated, it passes through periods of decalescence during which structural changes occur between the base metal (iron), carbon, and the alloying elements. The result of these changes is a solid solution, which means, in this instance, that both the iron and the alloy lose their separate identities and become a single material (*austenite*).

Rapid quenching prevents a change to the annealed condition but develops a structure called *martensite*. This structure has many small planes whose corners or edges project to the boundary of the grain and provide many small particles that produce maximum interference to slip, consequently developing a high degree of hardness. During this transformation stage, subzero temperatures (deep freeze) can be applied to ensure full hardness and promote uniformity of the steel.

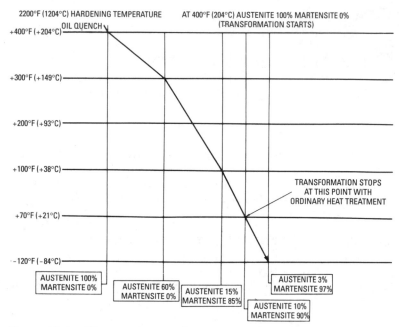

Figure 15-1 Diagram showing the percentage of austenite transformed to martensite in high-speed steel at temperatures between 400°F (204°C) and –120°F (–84°C).

In common heat-treating procedures, the decomposition of austenite to martensite stops when the temperature of the steel decreases to room temperature (Figure 15-1). By cooling at temperatures of –120°F (–84°C), it is possible to complete the decomposition of the retained austenite.

Cold-Treating Temperatures

The precise subzero temperatures for cold treating depends on the metal being treated, but it has been found that metals can be cold-treated between –100°F (–73°C) and –120°F (–84°C). The transformation of austenite to martensite stops at –150°F (–101°C); lower temperatures have no useful effect on high-speed steels. In addition, temperatures lower than –150°F (–101°C) can cause the metal to crack. Temperatures higher than –100°F (–73°C) are relatively ineffective in the cold-treating process.

The length of time required to cold-treat a metal depends on the material itself. It is necessary to chill high-speed steel thoroughly. Both ferrous and nonferrous metals (such as aluminum, cast iron, and steel) require a time cycle (alternately subjecting the parts to heat and cold, returning each time to room temperature before going to the other extreme) to stabilize the part completely. Although two to five cycles are recommended, the exact number of cycles depends on the stability required. The final cycle should be followed by temperatures of 200°F (93°C) to 300°F (149°C).

Convection Fluid

In any cold-treating procedure (such as stabilizing, hardening, or shrink-fitting) heat can be dissipated more rapidly if the unit is filled to two-thirds capacity with a convection fluid, such as methylene chloride or ethyl alcohol (denatured alcohol) M.P. 179°F (82°C), and the parts immersed in the fluid.

The parts should remain in the fluid at a temperature of –120°F (–84°C) for a sufficient length of time after dissipation of heat to ensure complete transformation within the material. The precise length of time will vary (for example, 42.5 pounds of steel contains 1000 Btu per hour). The rated capacity of the Deepfreeze Model F-120 is 1000 Btu per hour. Therefore, a period of 1 hour is required to dissipate the natural heat from the 42.5 pounds of steel, and an additional 1½ to 2 hours are required to complete the transformation of austenite to martensite.

Calculating Rate of Production

The following formula can be used to calculate the number of Btu that must be removed from a piece of metal:

$$H = WDS$$

in which H is the heat to be removed from material; W is the weight of material; D is the degrees temperature change desired; and S is the specific heat of material.

For example, how many long steel bars 4¼ inches in diameter × 4 inches in length can be chilled [temperature reduced from 80°F (26.7°C) to –120°F (–84°C)] per hour?

By referring to a table of standard weights, the weight of the material can be found to be 4.02 pounds per inch, or 16.07 pounds for a 4-inch bar (4.02 × 4). The specific heat of steel, as shown in Table 15-1, is 0.118, and the desired temperature change is 200°F (93°C). Substituting in the formula ($H = WDS$), you can determine the requirement to be removed from each piece:

$$H = (16.07 \times 200 \times 0.118)$$
$$= 379.252 \text{ Btu}$$

Table 15-1 Specific Heat of Common Materials

Material	Heat
Aluminum	0.212
Brass	0.092
Bronze	0.104
Cast iron	0.113
Copper	0.092
Glass	0.180
Marble	0.206
Nickel	0.109
Rubber, hard	0.339
Silver	0.056
Steel	0.118
Tantalum	0.033
Tin	0.054
Tungsten	0.034
Zinc	0.993

Since the Deepfreeze Cascade Model F-120 machine can remove 1000 Btu per hour, it is possible to handle

$$\frac{1000}{379.252} = 2.6368 \text{ pieces per hour}$$

Cold-Treating Procedures

Several procedures can be used in cold-treating the various types of steels to increase hardness by effecting a transformation of austenite to martensite. The beneficial effect of cold treating is lost unless heat treating is also properly performed in conjunction with cold treating.

High-Speed Tool Steel

Three procedures have proved satisfactory in the treatment of high-speed tool steels. It should be noted that Procedure No. 1 and Procedure No. 3 have similarities, but they differ with respect to the order in which the tempering and cold-treating operations occur. On the other hand, Procedure No. 2 has two cold-treating cycles. The three procedures are as follows:

- *Procedure No. 1*—Preheat at 1400°F (760°C) to 1500°F (816°C), depending on analysis [double preheating is recommended, using 700° to 1000°F (371° to 538°C) for the first preheating].

a. Heat to hardening temperatures, depending on analysis. [Test indicates unsatisfactory results of subzero treatment on high-speed steel if the hardening temperature is near the lower part of the range. Effective temperatures appear to range from 2300° to 2350°F (1260° to 1288°C).]

b. Quench in oil, lead, salt, or air.

c. Remove from quenching medium at approximately 200°F (93°C) and transfer to tempering temperature.

d. Temper 2 to 4 hours to hardness specification [1000°F (538°C) minimum].

e. Allow tool to cool to 150°F (66°C).

f. Cold treat to −120°F (−84°C) in Deepfreeze chilling machine for 3 to 6 hours, depending on cross section of material.

g. Allow tool to return to room temperature normally.

h. Repeat tempering cycle, using 25°F lower temperature for 2 to 4 hours.

• *Procedure No. 2*—Preheat as in first procedure to 1400° to 1550°F (760° to 843°C) (double preheat whenever possible).

a. Heat to hardening temperatures of 2300° to 2350°F (1260° to 1288°C), depending on analysis.

b. Quench in oil, lead, salt, or air.

c. Tool can cool in air if removed from quenching medium at higher temperatures to 150°F (66°C).

d. Transfer to Deepfreeze unit at −120°F (−84°C) for 3 to 6 hours, depending on cross section.

e. Allow to return to room temperature.

f. Temper to specified hardness.

g. Transfer to Deepfreeze unit at −120°F(−84°C) when tool has cooled at approximately 150°F (66°C) from the tempering temperature.

h. Remove from Deepfreeze unit and allow tool to return to room temperature.

i. Retemper at 25°F lower than original tempering temperature for 2 to 4 hours.

- *Procedure No. 3*—This is the same as Procedure No. 2 except that the second subzero cooling is omitted. The cycle consists of hardening, subzero cooling, and double tempering. Note that Procedure No. 1 has only one subzero cooling, but it follows the first tempering operation.

Sometimes cutters and tools that have been heat-treated in the usual manner (without subzero cooling) and finished by grinding can be improved by being subjected to the subzero treatment. Hardening and tempering tends to stabilize whatever austenite is retained. Therefore, the subzero treatment of stock tools will not increase the tool durability to the same degree as when this treatment is included in the heat-treating process. Double-tempered tools show even less improvement with this after-finish subzero treatment.

If precise dimensional qualities are desired, the tool should be subjected to subzero temperature before it is finish ground. The reason for doing this is that in accomplishing the transformation of the retained austenite to martensite, a slight increase in the size of the tool results because martensite is larger in volume than austenite. Therefore, if finished stock tools are taken from the tool crib and subjected to a subzero treatment, those requiring absolute dimensional stability should be measured both before and after treatment to determine whether the original dimensions have changed.

If a slight increase in the size of the tool results from cold treating, the tool can usually be returned to correct size by grinding. Hardness of cutting tools can increase perceptibly with this treatment if the original tempering operation was incomplete. If this increase in hardness does occur, the tools can be drawn back to the usual hardness working ranges.

High-Carbon Steel

The amount of austenite retained in ordinary carbon steel is usually not large enough to be detrimental. However, in some instances where the machine part is to be placed in extremely hard usage, it is necessary to complete the transformation of austenite by means of subzero temperatures. The cupping punch that is used in drawing out steel cartridges is a typical example. Formerly, only 3000 shells could be drawn out with a punch, but after cold treating, 30,000 shells could be drawn with a punch.

For another example, the tool life of step-cut reamers was increased from 100 to 400 holes by subjecting the reamers to a temperature of –120°F (–84°C) for a period of 2 to 4 hours. The flute ends also wear evenly rather than becoming tapered.

Stabilizing Dimensions

The transformation of austenite may occur naturally over a period of months or years. This transformation into martensite is accompanied by an increase in volume. Such changes may be serious in the case of precision gages, close-fitting machine parts, and so on. The subzero treatment has proved effective in preventing such changes. Gage blocks, for example, are stabilized by hardening followed by repeated cycles (two to six cycles are common) of chilling and tempering. This transforms a large percentage of the austenite into martensite.

Subzero treatment will always cause an increase in size. Machine parts subjected to repeated and drastic changes in temperature (such as in aircraft) may eventually cause trouble. As the austenite gradually changes to martensite, the pieces grow or warp and may cause a seizure or some other type of failure. The practical remedy is to apply the subzero treatment before the final machining operation.

Subzero Chilling

Subzero chilling, another aspect of the cold-treating process, is used in industry to:

- Shrink metals for fitting and assembling.
- Test aircraft instruments and materials.

One of the chief advantages of chilling parts for shrink-fit assembly is that there is less possibility of altering the characteristics of the metal by chilling than by heating. Tools or parts that have sections or portions that differ in size are less likely to be distorted when the chilling method is used. Chilled parts are also easier to handle on the assembly floor than heated parts. The hazard of oxidation is eliminated on chilled parts, and a finishing operation is not required to remove the oxidation from the surface of the metal.

Subzero chilling is used extensively to test aircraft instruments and materials. It lends itself to testing instruments and materials that normally function in the low temperatures encountered at high altitudes by aircraft.

The use of subzero temperatures in testing aircraft parts has resulted in many improved developments in metals, plastics, rubber, and other materials that have proved invaluable in modern production. It is usually comparatively simple to adapt the chilling unit for testing these materials to subzero temperatures.

Summary

A cold-treating process is used as a supplement to heat treating in hardening metals. Cold-treating machines are built for high-production work and are standard equipment in plants that use the cold-treating process.

As steel is heated, it passes through periods of decalescence during which structural changes occur between the base metals. The result of these changes is a solid solution, which means, in this instance, that both the iron and the alloy lose their separate identities and become a single material called austenite. Rapid cool-off prevents a change to the annealed condition but develops a structure called martensite.

In common heat-treating procedures, the decomposition of austenite to martensite stops when the temperature of the steel decreases to room temperature. The precise subzero temperatures for cold treating depend on the metal being treated, but it has been found that metals can be cold-treated between –100°F (–73°C) and –120°F (–84°C). The transformation of austenite to martensite stops at –150°F (–101°C); lower temperatures have no useful effect on high-speed steels. In addition, temperatures lower than –150°F (–101°C) can cause metal to crack.

Several procedures can be used in cold-treating the various types of steels to increase hardness. The beneficial effect of cold treating is lost unless heat treating is properly performed in conjunction with cold treating.

Subzero treatment can transform a large percentage of the austenite into martensite. This, in turn, causes the tool or machine part to become stabilized in size. Thus, the treatment is used for precision gages and parts that are subjected to large temperature changes.

Subzero chilling is also used in industry to shrink metals for fitting and assembling and to test aircraft instruments and materials. Improved development in metals, rubber, plastic, and other materials has proved invaluable through the use of subzero temperature testing.

Review Questions

1. What is austenite?
2. Explain martensite.
3. At what temperature will metal generally crack?
4. What are the temperatures for the cold-treating process?
5. Explain the need for subzero treatment for precision gages.

6. Why does metal increase in size when subjected to cold treating?

7. Why is subzero chilling used in industry?

8. What products have been improved through a chilling process?

9. Describe hardness in terms of the slip-interference theory.

10. What is martensite?

11. Define *decalescence*.

12. What is the range of temperatures for cold treating?

13. Why is a convection fluid used in cold treating?

14. What is another name for ethyl alcohol?

15. What is the formula used to calculate the rate of production?

16. What is the specific heat for these metals: aluminum, brass, silver, zinc, tin?

17. What are the three procedures that have proved satisfactory in the treatment of high-speed tool steels?

18. Subzero treatment will always cause an increase in _____.

19. The transformation of _____ may occur naturally over a period of months or years.

20. Austenite gradually changes to _____.

Chapter 16

Automatic Lathes

Automatic lathes are ideally suited for medium to long production runs. In many instances, the automatic cycle enables the operator either to handle additional units or to perform other work.

Automatic Turret Lathes

The automatic turret lathe is especially suited to small-lot and long production runs (Figure 16-1). All functions of this lathe are under complete computer numerical control so that the primary functions of the operator consist of loading the machine and inspecting the finished part. The lathe is a universal turning center designed for bar, chucking, and shaft work.

Figure 16-2 shows many different parts in a variety of shapes, sizes, and types of material. The lathes have one thing in common. They can perform several operations to produce uniform and accurate parts in large quantities with little scrap.

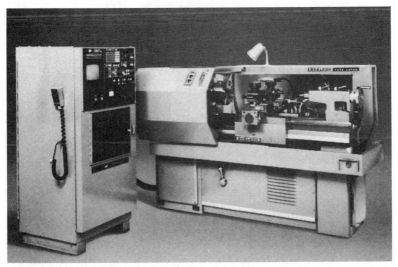

Figure 16-1 This computer numerical-controlled lathe is a universal turning center designed for bar, chucking, and shaft work.
(Courtesy Sheldon Machine Co.)

Figure 16-2   Many different shapes and sizes of parts produced from different types of materials. *(Courtesy Sheldon Machine Co.)*

To get from "print to finished" part, a number of steps must take place (Figure 16-3 and Figure 16-4). The control system enables the operator to increase work efficiency. The control system also has a computer retrieval terminal to display operating programs, tool compensation, commands, positions, and program edits (Figure 16-5). The automatic turret lathes also have a number of options to increase the capability of the lathe (Figure 16-6 and Figure 16-7).

Automatic Threading Lathes

All types of internal and external threads can be produced on the automatic threading lathe (Figure 16-8). These machines can produce

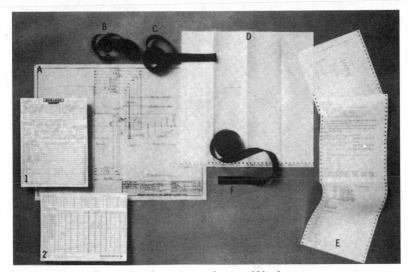

Figure 16-3   From the dimensioned print (A), the programmer obtains necessary information to fill in forms 1 and 2. The first form allows for general comments and tooling, while the second contains the numerical data describing the rough stock and the finished part. These forms are then given to a typist who, with the use of master tape (B), fills in the information written by the programmer to produce input tape (C). This tape is transmitted to Sheldon by dataphone and processed by computer. With this information, the computer generates the complete machining cycle. It provides a plot showing the paths of each tool (D); complete printout, including the time to produce the part (E); and the final output tape (F). This tape contains the setup instructions and operating program and is transmitted to the user by a dataphone terminal.

(Courtesy Sheldon Machine Co.)

threads in much less time than threads can be produced either by thread grinding or by thread milling. The supporting saddle of the automatic threading lathe is moved along the bed to position the tool in relation to the work. The saddle is clamped during threading, which is performed by a longitudinal slide and cross slide mounted on the saddle. The slides are cam controlled and provide a very accurate thread at a high rate of production.

Figure 16-4 Tape containing the setup instructions and operation program is transmitted by dataphone terminal. *(Courtesy Sheldon Machine Co.)*

Figure 16-5 These controls give the operator complete information about the status of the lathe and the machining operation.
(Courtesy Sheldon Machine Co.)

Figure 16-6 Chuck with collet pads shows bar feed application. Simply by changing jaws, the lathe can be converted to either a bar or chucking application. *(Courtesy Sheldon Machine Co.)*

Figure 16-7 Optional eight-station automatic indexing front turret permits total of 12 tools to be used in one setup. This helps eliminate tool changes from job to job. This turret uses standard qualified 1″ × 1″ × 4″ tools. *(Courtesy Sheldon Machine Co.)*

Figure 16-8 Automatic threading lathe. *(Courtesy Gisholt Machine Co.)*

Summary

For long production runs, the automatic turret lathe is an ideal piece of machinery. As an example, the functions of the lathe are under complete computer control so that the primary functions of the operator consist of loading the machine and inspecting the finished part.

The automatic turret lathe is capable of producing parts in a variety of sizes, shapes, and types of material. They can perform several operations to produce uniform and accurate parts in large quantities with little scrap. All types of internal and external threads can be produced on the automatic threading lathe in much less time than can be produced either by thread grinding or by thread milling.

Review Questions

1. What is an automatic turret lathe?
2. All types of internal and external _____ can be produced on the automatic threading lathe.
3. Why are automatic lathes important in long production runs?
4. What are some of the advantages of the automatic threading machine?
5. Where are the setup instructions and operation program located?
6. The computer numerical-controlled lathe is a universal turning center designed for bar, chucking, and _____ work.

7. All types of internal and external _____ can be produced on the automatic threading lathe.

8. What is a saddle?

9. What does "print to finish" mean?

10. All functions of the automatic turret lathe are under complete _____ numerical control.

Chapter 17

The Automatic Screw Machine

Because of modern developments, the term *screw machine* is practically a misnomer. The original screw machine was developed many years ago. Development of the original screw machine, by the addition of automatic devices, has resulted in machines being designed for various applications with each having its own field.

The screw machine was designed originally for making small screws and studs. The flexibility of the machine has adapted it to a large variety of work, resulting in specialized types of machines adapted to other operations.

A screw machine is a type of turret lathe designed in such a way that it rapidly positions a series of tools to the piece of stock held in the chuck for a series of operations to be performed on it. The screw machine differs from the lathe in that it is provided with a front and rear cross slide, and the tailstock is replaced by a slide with a turret mounted on it. The turret is provided with several tool positions, which can be indexed into a working position as the slide is advanced, permitting each successive tool to perform an operation on the part being turned.

Classification

The numerous types of screw machines can be classified as follows:

- Type of operation—Plain or hand operated; semiautomatic; automatic.
- Type of spindle—Single; multispindle.

Drilling, boring, reaming, forming, facing, and so on are examples of operations that can be performed on the screw machine. In a semiautomatic screw machine, the rough piece of work must be placed in the chuck by the operator. The various operations (including rotation of the turret to bring the tools into position) are performed automatically.

In the automatic screw machine, the rough casting or drop forging is placed in a hopper or magazine. Then, they pass to the chuck, which grips them automatically for the cycle of machining operations. The automatic cycle can be compared to that of a boiler and automatic stoker in which the only attention required of the operator is to keep a supply of coal available.

A single-spindle screw machine feeds a single piece of stock that is machined by each tool in succession in the indexing turret. A multiple-spindle machine feeds a number of pieces of stock progressively, that is, the spindles index so that the stock in each spindle is fed progressively to each tool. The indexing operation is performed automatically on semiautomatic and automatic machines. The indexing turret is one in which an appropriate gear is turned through equal angles to lock and bring each tool into working position.

Operating Principles

The general principles of automatic screw machines are relatively simple. The single-spindle machine is referred to here as an example.

In the machine shown in Figure 17-1, the length of bar stock feed is preset by a simple hand-crank adjustment and is indicated on a scale bar. A feed shell, attached to a feed tube, is used to grip the bar stock. The feed tube is mounted in a cartridge and attached to the feeding member through a latch. Thus, a quick method for changing feed shell pads is provided (Figure 17-2). The feeding member is operated by a cam mounted on the main camshaft.

Figure 17-1 A single-spindle automatic screw machine. *(Courtesy Cleveland Automatic Machine Co.)*

Standard cams control the forward and return motions of the turret as well as that of the two cross slides and the independent cutoff attachment. Turret feeds can be varied by adjusting standard

Figure 17-2   Collet pads can be changed without removing the master collet from the chuck. *(Courtesy Cleveland Automatic Machine Co.)*

cams on the regulating wheel. Cross-slide feeds are a proportion of the turret feeds. The cross-slide cams can be set relative to any turret operation by adjusting the cross-slide cam drum on the main camshaft (Figure 17-3).

Cutting lubricant must be used frequently to avoid excessive heat and to prevent ruining either the work or the tools. The operating parts should be kept clean and free of any gummy coating of dried lubricant, chips, or dirt. If this is neglected, the machine can become clogged to an extent that will interfere with its proper operation. The machine must be well cared for since exacting demands are made on the machine, and it is expected to run continuously and work within extremely close limits.

Selection and Use of Tools

The quality of work produced by the automatic screw machine is determined by the proper selection and use of tools. These factors also are related to the accuracy, finish, and speed with which the work is turned out. The tools also must be properly ground and correctly set for best results.

When setting up a job, experience in "tooling up" is advantageous and of great help on either special jobs or very difficult jobs.

Figure 17-3 Front and rear cross slides. Cross-slide tools can be adjusted precisely in relation to turret tools. *(Courtesy Cleveland Automatic Machine Co.)*

However, tool principles are relatively simple. Even if the machinist is unfamiliar with a particular type of machine, he or she should be able to obtain the desired results by combining proper care with studying the instructions.

Types of Tools
Many different kinds of tools are used because a wide variety of work is performed on automatic screw machines. Usually, selection of the type of tool for an operation is governed by experience, although certain principles and suggestions can be followed (Figure 17-4).

External Turning Tools
Balance turning tools, plain hollow mills, adjustable hollow mills, box tools, swing tools, and knee tools are all used in the turret. Suitable tools for operations on the end of the work are centering and facing tools, pointing tools, and pointing-tool holders with circular tools.

Internal Cutting Tools
Drills, counterbores, and reamers are held in drill or floating holders. Recessing swing tools are also used for internal operations.

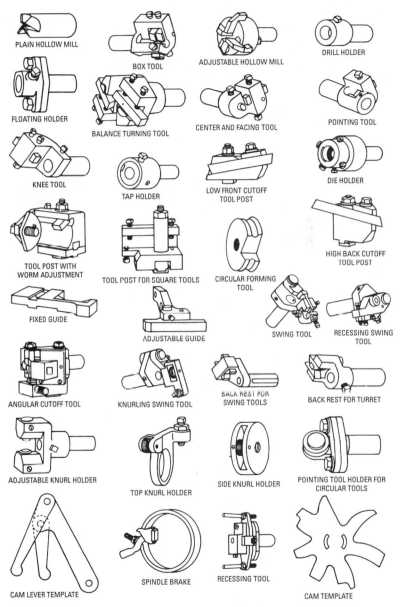

PLAIN HOLLOW MILL

BOX TOOL

ADJUSTABLE HOLLOW MILL

DRILL HOLDER

FLOATING HOLDER

BALANCE TURNING TOOL

CENTER AND FACING TOOL

POINTING TOOL

KNEE TOOL

TAP HOLDER

LOW FRONT CUTOFF TOOL POST

DIE HOLDER

TOOL POST WITH WORM ADJUSTMENT

TOOL POST FOR SQUARE TOOLS

CIRCULAR FORMING TOOL

HIGH BACK CUTOFF TOOL POST

FIXED GUIDE

ADJUSTABLE GUIDE

SWING TOOL

RECESSING SWING TOOL

ANGULAR CUTOFF TOOL

KNURLING SWING TOOL

BACK REST FOR SWING TOOLS

BACK REST FOR TURRET

ADJUSTABLE KNURL HOLDER

TOP KNURL HOLDER

SIDE KNURL HOLDER

POINTING TOOL HOLDER FOR CIRCULAR TOOLS

CAM LEVER TEMPLATE

SPINDLE BRAKE

RECESSING TOOL

CAM TEMPLATE

Figure 17-4 Automatic screw machine tools and accessories.

Threading Tools

Taps and dies, chasers mounted in opening die holders, and thread rolls held in cross-slide knurl holders or in knurling swing tools are the threading tools most generally used.

Knurling Tools

Top and side knurl holders are used on the cross slides. Adjustable knurl holders and knurling swing tools are used for knurling from the turret.

Forming and Cutoff Tools

Circular forming tools and cutoff tools are used on the regular cross-slide tool posts with a worm adjustment. Square tools and thin straight-blade cutoff tools are used on the cross slides. The angular cutoff tools are used in the turret.

Supporting and Auxiliary Tools

Swing tools, fixed and adjustable guides for operating swing tools, backrests for turrets, and spindle brakes are other equipment that can be used on automatic screw machines.

General Suggestions for Tool Selection

The balance turning tool is usually for straight rough turning with plain hollow mills and adjustable hollow mills as the next choices. The box tools are preferable for straight finish turning. However, adjustable hollow mills are sometimes used. The knee tool is used for turning scale on machine steel. A pointing tool of the box type is preferred for straight turning on long and slender work as the bushing ahead of the blade holds the work steady.

The swing tool is used for taper turning and for some form turning where the curve is long, shallow, and continuous. Circular form tools are used on the cross slides for all form turning and for straight turning behind shoulders. Generally, they are used wherever they can be applied, and the length of cut is not great in proportion to the diameter. Extra-wide circular form tools are held in the tool post with worm adjustment. The swing tool can be used if the turning cannot be accomplished otherwise. Form turning can be performed occasionally with a straight-bladed tool held in the tool post for square tools. However, this is seldom used since the tool has no advantage over the circular form tool, and it does not last as long as the circular tool.

The center drill clamped in a floating holder is usually used for centering the end to be drilled, as this setup permits easy adjustment. If the end of a bar requires facing, as well as centering, a centering and facing tool that combines the two operations can be used.

To round off burr, or point the end of work, a pointing tool holder with a circular tool should be used if the work is strong enough to prevent springing during the operation. If the work is long and slender, use a pointing tool of the box type, which supports the work just ahead of the cut.

Drills, reamers, and counterbores should be mounted in a floating holder so that they can be easily adjusted centrally with the work. A drill holder is not used unless the tool is so fine that it will spring enough to correct any inaccuracy in setting. A groove or recess inside the piece can be cut with the recessing swing tool.

Taps and dies should be held in non-releasing tap and die holders. The releasing type should be used only in instances where it is not possible to develop a thread lobe on the cam because of its slow movement, and the tool must dwell before backing out.

If a thread is to be rolled, the top knurl holder for the cross slide is often used for carrying the thread roll. The cutoff operation immediately follows the threading operation on the same tool post. If the thread rolling operation is heavy, the side knurl holder is preferred for the cross slide as it is free from any tendency to pull up on the cross slide. If there is no room on the cross-slide posts for the holders to carry a thread roll or if the thread is to be placed in such a position that tools on the cross slide cannot be used, a thread roll in a swing tool can sometimes be used. Frequently, it is advantageous to roll a short thread on brass, aluminum, or other soft metals where the thread cannot be cut with a die. The opening die holder is used in place of the button die when the spindle is run in only one direction.

The adjustable knurl holder is preferred for knurling on straight work because it carries two knurls that swivel for straight or diamond knurling, and it gives a balanced cut. If the part to be knurled is not straight but is formed (convex for example), the top or side knurl holders are used. This also applies if the knurling is behind a shoulder. The knurling swing tool is applied only on formed knurling or behind shoulders when the cross-slide holder is not available. The top knurl holder is usually preferable to the side knurl holder because it permits a cutoff tool to be used on the same tool post with it.

For straight cutoff work where burring, rounding, or forming of the bar or piece cut off is not required, a thin cutoff blade of the straight type held in a cutoff tool post is used. The circular cutoff tool is used most frequently because it can be shaped readily to part the piece properly from the bar, performing a chamfering or similar operation at the same time. Once formed, the work retains its shape after repeated grindings.

If the cutoff or back end of the work is to be pointed or coned, the angular cutoff tool should be used. The fixed guide is intended only as a pusher for forcing the angular cutoff tools, as well as the various swing tools, when they are not fed along the piece against the work. If a swing tool is to be fed along the piece for turning after being forced into position by the guide, the adjustable guide should be used. This also applies to straight, taper, or formed work because adjustment of the guide can compensate for any slight error in the tool or machine.

In certain instances where a cut requires considerable side pressure that tends to spring the work (as in thread rolling, knurling, or forming operations), a backrest should be used in the turret for support if the work permits. Swing-tool cuts on slender work occasionally require the use of a backrest in the swing-tool shank. Backrests should not be used unless they are really necessary.

Setting Up an Automatic Screw Machine

The tooling and setup preparations given here are relative to the Brown & Sharpe No. 2 automatic (high-speed) screw machine. Instructions given here are for the operator who has been furnished the cams, necessary tools, and a drawing. The method of setup is practically the same as for other machines. Some types are equipped with a speed-change mechanism that requires adjusting an additional set of dogs, but the general outline of operations is the same. Ingenuity and mastery of the technique are essential in tool selection and setup to realize the greatest measure of economy and efficiency that is possible on high-production automatic screw machines.

The general sequence can be followed in practically all instances, but familiarity can develop slightly different ways of making the various settings and adjustments. It will be noted that the tools are set up and adjusted in the same general order of their operations as that laid out for the job. For the particular job to be used here as an example, the data regarding dimensions of the piece, outlines of the cams, speeds, and so on, are shown in Figure 17-5. The outlines of the three cams are superimposed on each other; drawing the cams in this manner makes the sequence or order of operations more clear at a glance.

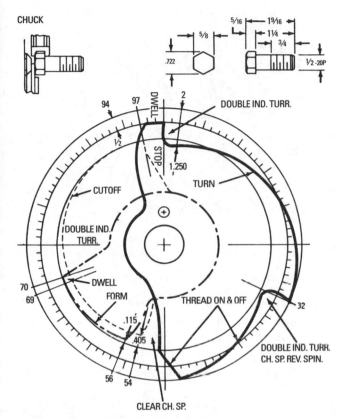

Figure 17-5 Drawing of the piece and cams with the order of operations. The job data are given as follows: double-index turret; No. 22F left-hand box tool; double-index turret and reverse spindle; thread off; clear and double-index turret; form—back slide; cutoff—front slide; feed stock to stop (4 inches). Spindle speed: 165 rpm forward, 765 rpm reverse. Driving shaft speed, 240 rpm.

It is a good practice to list the order of operations with the number of spindle revolutions and feeds required for each operation and the kind and size of tools that are to be used. If the machine has been used previously for a different job, it is necessary to remove the tools and cams and arrange the machine for either single- or double-indexing (Figure 17-6).

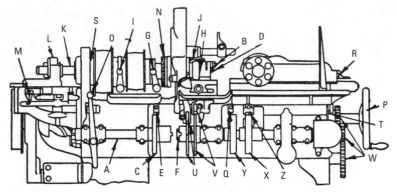

Figure 17-6 Diagram of Brown & Sharpe No. 2 automatic screw machine.

The machine is set with the chuck open so that the grip of the new collet can be adjusted and checked as follows:

1. Raise the trip lever Q (see Figure 17-6).
2. Turn the handwheel P until the chuck is open, lift latch L, and pull out the feed tube K.
3. Take out the feeding finger, which is threaded into the tube (left-hand thread), and replace with the feeding finger to be used on the new job.
4. Replace the feed tube, but do not leave the latch L down.
5. Unscrew the chuck nut on the end of the spindle with a pinwrench, and the spring collet will slip out.
6. Do not disturb nut J as it is used to adjust the spindle box.
7. Install a new spring chuck, and before turning on the chuck nut, make sure that the sleeve into which the collet is inserted is completely back in place. This sleeve, like the collet, is free when the chuck nut is removed and is sometimes partly pulled out with the collet.
8. There is a slot in the back end of the sleeve that fits over a pin inside the spindle; turn the chuck nut tightly against the spindle nose.

The grip of the spinning chuck is adjusted by opening and closing the chuck with a hand lever inserted in the chuck form at G. The grip of the collet on the stock is regulated by means of the knurled nut N (see Figure 17-6). On the No. 00 and No. 0 machine sizes, the chuck operating mechanism is at the extreme left-hand end of the spindle, and the two nuts on the end of the spindle provide for regulating the grip of the chuck. The grip of the collet should be firm enough that it will not let the work slip but not so tight that the levers that operate the chuck will be broken.

Arrangement of Belts for Correct Spindle Speed

In Figure 17-5, the specified spindle speeds are 165 rpm for forward speed and 765 rpm for reverse speed. The diagram (Figure 17-7) shows the right-hand spindle-driving belt on the large pulley and the left-hand spindle-driving belt on the small pulley. The first countershaft should run at 215 rpm with the belt on the third cone pulley.

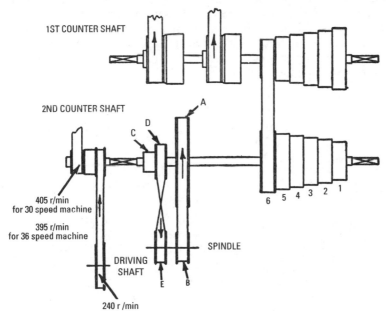

Figure 17-7   Belt diagram for Brown & Sharpe No. 2 automatic screw machine (high-speed).

Slow forward speed for threading on, and fast reverse speed for threading off, should be arranged for threading jobs requiring reversal of the spindle. This same reverse speed is used for cutting operations using left-hand tools.

The spindle speeds are changed on motor-driven machines by means of change gears and spindle pulleys (sprocket on No. 2G machine) located in the base of the machine. A switch is located at the right-hand end of the tank table for reversing the motor as required. After reversing the motor, make certain that the backshaft driving belt is running in the correct direction.

Indexing the Turret

Indexing is performed by a simple Geneva movement, and the turret is locked by means of a taper bolt that is withdrawn during the index movement by a cam connected with the turret change mechanism. The two kinds of indexing are single and double.

Figure 17-8 shows the position of turret change rolls and turret locking pin cam for both single and double indexing. One roll on the turret change gear with the proper cam for withdrawing the turret locking pin is used for single indexing. Two rolls (using only alternate holes in the turret), with an additional cam surface to withhold the turret locking pin the required amount of time, are used for double indexing.

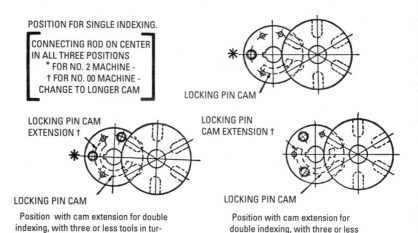

POSITION FOR SINGLE INDEXING.

CONNECTING ROD ON CENTER IN ALL THREE POSITIONS
 * FOR NO. 2 MACHINE -
 † FOR NO. 00 MACHINE -
CHANGE TO LONGER CAM

LOCKING PIN CAM

LOCKING PIN CAM EXTENSION †

LOCKING PIN CAM EXTENSION †

LOCKING PIN CAM

Position with cam extension for double indexing, with three or less tools in turret and with positive pull-off to the full turning capacity as above, but no push-on.

LOCKING PIN CAM

Position with cam extension for double indexing, with three or less tools in turret and with positive pull-off and push-on equally divided, but for short lengths only.

(A) Position for single indexing.

(B) Position for double indexing.

Figure 17-8   Diagram showing positions of turret change rolls and turret locking pin cam for single and double indexing.

Changing from Double to Single Index

The method of changing from a double to a single index is similar on all three sizes of machines, except that the turret locking pin cams are in slightly different positions. Remove the cover from the turret slide, and remove the turret change roll. This is the roll not aligned with the eccentric pin on the turret change shaft. It is mounted on a stud held in place by a wedge pin that can be driven out with a steel punch and hammer. The procedure is as follows:

1. Remove the section of the cam that is square and without rise as this is the part of the cam that holds the turret locking pin out of the turret during indexing of the second hole.

2. Trip the lever, and turn the handwheel to the completion of one index. Always try the mechanism by hand to avoid breaking any parts if the change has not been made correctly.

3. Replace the turret slide cover and proceed with the setup.

The change gears are replaced by taking off the change gears at W (see Figure 17-6), which were used for the previous job, and replacing them with the gears required for the new job. These gears are specified on the drawing (see Figure 17-5). Then, arrange as shown in Figure 17-9.

Setting Cross-Slide Tools

When the top of the work revolves toward the cross-slide post, a rising block (D in Figure 17-10) is used under the post to bring the tool to the proper height. Loosely clamp the circular cutoff tool to the proper cross-slide tool post by the bolt B, and hook bolt A, setting the cutting edge as close as possible to the height of the center of the work.

The cutting edge of the circular tool is adjusted to the work by turning nut E that, by means of an eccentric, tilts the thin plate F to which the tool is clamped. The clamp screw K on the post of the No. 2 size of machine keeps the eccentric from turning and should be loosened before adjustment. When the tool is set correctly, clamp it securely by means of bolts A and B, as shown in Figure 17-10. (See also Figure 17-11.)

Adjusting the Cutting Tool to Proper Distance from Chuck

It is the most desirable practice to work as close to the chuck as possible with all tools. The cutoff tool is the one that governs the distance from the chuck more often than the others. The cutoff tool

CAM DESIGN WORK SHEET

PART NAME OR NO...... Example 1 .... MATERIAL ...... BRASS......................

SURFACE FEET FOR STOCK............
" " " THREAD ....
" " " DRILL ....
" " " TURN.............

SPINDLE R P M { FORWARD ... 1200 ....
BACKWARD ....

SECONDS...30 ....
GROSS PRODUCTION PER HOUR ....120 ....
MADE ON ... No. 2 ....

Order of Operations	Throw	Feed	Spindle Revolutions: For each Operation	Spindle Revolutions: After Deducting for Operations Overlapped	Spindle Revolutions: Readjusted to Equal Revolutions Obtainable with Regular Change Gears	Hundredths: For Spindle Revolutions in Preceding Column	Hundredths: For Operations That are Overlapped
Feed Stock to Stop				24	24	4	
Index Turret				24	24	4	
Rough Turn- No. 22 Balance Turning Tool.........	1.500	.010	150	150	150	25	
Index Turret................				24	24	4	
Finish Turn - No. 22B Box Tool..........................	1.500	.010	150	150	150	25	
Clear............................					24	4	3
Form - Front Slide ....{	.182	.003	60				10
	.010	.001	18				3
Index Turret 4 Times							16
Cut Off - Back Slide	.574	.0033	175	175	204	34	
96 Hundredths equal				547			
Estimated Total Revs., if spindle runs 1200 R P M Continually.......................				570			
Nearest Actual Spindle Rev. available on Machine					600	100	

NOTE: A dimensioned pencil sketch of the piece is often drawn in blank space at top of this sheet.

Figure 17-9 Cam design worksheet.

is set by loosening the nut G in Figure 17-10 and sliding the tool post either toward or away from the face of the chuck. In some instances, it is the form tool (and sometimes it is a turret tool) that governs the distance to set the cutoff tool from the chuck face, but a study of each job and the tools to be used will enable the operator to determine the proper distance.

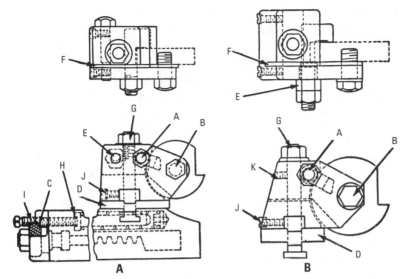

Figure 17-10   Clamping and adjustment of cross-slide circular tools.

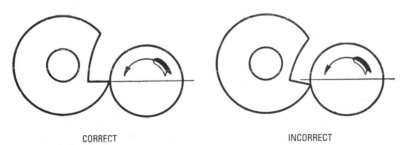

CORRECT   INCORRECT

Figure 17-11   Correct (left) and incorrect (right) methods of setting the circular tool with reference to the work.

Adjust the Form Tool to Line Up with the Cutoff Tool

The stock should be fed out far enough for the cutoff tool to cut off a thin piece of stock. Start the spindle, place the hand lever in the hole in the cam lever that operates the cross slide on which the cutoff tool is mounted, and raise the lever until the tool slightly nicks the bar. Then, stop the spindle and bring the form tool up in the same manner. Align it with the groove cut by the cutoff tool, according to the way in which the job was laid out.

Placing the Cams

Preliminary to this operation, draw the cross slides back by means of the nut C in Figure 17-6. Care should be taken not to place the cams on backward. The upper side of the camshafts are "coming."

Adjusting the Cutoff Tool to the Cam Lobe

Turn the handwheel P in Figure 17-6 until the cam lever operating the cross slide that carries the cutoff tool is at the highest point of the cam lobe and about to drop back. Turn the nut C in Figure 17-10 with a wrench, feeding the cutoff tool inward until the piece is cut off and the tool has traveled far enough past center to leave the end of the bar smooth. Turn the handwheel P in Figure 17-6 until the roll drops off the high point of the cross-slide cam and the cutoff tool withdraws until it clears the diameter of the stock.

Adjusting the Turret to the Correct Distance from the Chuck

With the lead lever roll on top of the cam stop lobe, measure with a rule from the face of the chuck to the turret, and adjust the turret slide by means of a screw R in Figure 17-6 until the turret is at the required distance specified on the cam drawing. On cams designed for operating the swing stop, the turret adjustment figure is given from the top of another convenient lobe.

Setting the Stock for Length

Without changing the setting of the machine, place the stock stop in one of the turret holes, hold out the turret locking pin, and revolve the turret until the stop is in line with the spindle. Move the cutoff tool toward the center of the work with the hand lever of the cross slide and then measure from the front edge of the cutoff tool a distance equal to the length of the piece. Set the stock stop to this point, clamp it securely (see Figure 17-6), and turn the handwheel until the swing stop has reached full downward position. Then, adjust the swing stop arm to the center of the stock and to the proper distance from the cutoff tool.

Setting the Chuck and Feed Trip Dog

The dog is fastened to the left-hand side of the carrier Y in Figure 17-6 and operates the trip lever Q. The dog should be set to trip the lever just as the cutoff tool clears the diameter of the stock. Sometimes, in close timing of jobs, the lever is tripped just before the cutoff tool clears the diameter of the stock. This is possible because the chuck has to open before the stock feeds. By the time the chuck is open, the cutoff tool is out of the way.

Setting Turret Indexing Trip Dogs

Turn the handwheel P in Figure 17-6 until the lead lever roll T begins to drop off the stop lobe of the cam. Then, set one of the trip dogs on either side of the carrier X so that it trips the lever Z and sets the indexing mechanism in operation. Turn the handwheel, and observe that the indexing of the turret consists of three movements:

1. The turret slide withdraws.
2. The turret then revolves from one station to the next.
3. The turret slide advances into position again, where its farther forward movement is controlled by the outline of the lead cam.

When the turret slide has advanced into position, the lead roll lever should be at the bottom of its drop on the cam, as shown at point A in Figure 17-12. If the roll has dropped only a portion of the distance, as at point B, the following turret must be "jabbed" into the end of the work when the slide advances into position. This means that the trip dog should be moved to trip the indexing mechanism slightly later. The easiest way to tell when the roll begins to drop off a cam lobe is to feel the motion of the turret slide by pressing the thumb against the back of the slide and its bearing on the machine bed, while turning the handwheel.

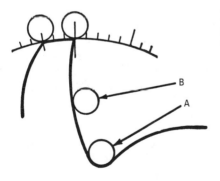

Figure 17-12 Diagram showing position of lead lever roll at beginning (A) and at end (B) of indexing.

After setting the trip dog for the first index point, turn the handwheel until the roll begins to drop off the high point of the next lobe, and set the indexing trip dogs until the index point just prior to the thread lobe is reached. When setting the dog at this point, the spindle reverse trip dog should also be set (Figure 17-13).

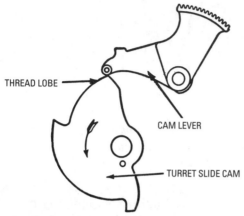

THREAD LOBE

CAM LEVER

TURRET SLIDE CAM

Figure 17-13 Position of lead lever roll on thread lobe when the spindle reverses.

Setting the Spindle Reverse Trip Dog

Turn the handwheel until the roll just begins to drop off the high point of the thread lobe, which can be felt at the joint of the turret slide and machine bed as described previously. At this point, set the trip dog on the side of the carrier C in Figure 17-6 to operate the trip lever E and reverse the spindle.

Setting the Indexing Trip Dogs

If threading is the last turret operation, set the necessary number of trip dogs required to rotate the turret so that the stock stop is in position when the stop lobe on the cam comes from under the lever roll. If only three tools are to be used, change to double indexing, and place the tools in alternate turret holes. Use three dogs on the turret dog carrier.

Adjusting the Feed Slide for Length of Stock

The crank M on the feed slide in Figure 17-6 provides a means of changing the length of stock fed whenever the stock-feeding mechanism is in operation, and a scale on the feed slide bracket shows the approximate length fed. The feed slide should be set to allow not more than $\frac{3}{8}$ inch in excess of the desired length. The stock should be set to feed out to the stock stop, which has already been set. After setting the feed slide, throw in the latch L in Figure 17-6, start the spindle, and throw in the starting lever O.

Placing and Adjusting the First Turret Tool

Throw out the lever O (see Figure 17-6) when the roll drops off the first stop lobe, and let the spindle continue to run. Place the first tool in the turret hole next to the stop. Set it well back into the turret, and clamp it securely. Place the hand lever for operating the turret in position, and feed the tool a short distance onto the work. The spindle can then be stopped, the turned diameter measured, and any adjustment that is necessary made on the tool to obtain the correct size.

Then, throw in the starting lever, or turn the handwheel until the roll is at the high point of the lobe, and stop at that point. The tool is fed onto the work by means of the hand lever, continuing until the correct length is turned. Adjustment should be made, loosening the bolt that clamps the tool in the turret and adjusting the tool outward to the proper length. Set the other tools by the same method, except threading tools, which should be left until the other tools have been set and the correct sample piece is obtained.

Adjusting the Form Tool

This tool is the last to work on the piece prior to cutting off. Therefore, it is the last tool to be adjusted. On jobs where the forming operation takes place earlier, the tool should be adjusted as soon as the tools preceding it have been set.

The form tool should be adjusted with the lever roll on the highest point or dwell of the cross-slide cam. The form tool can be adjusted to cut to proper depth by means of the nut C in Figure 17-10. The nut should be clamped by tightening the setscrew as soon as the correct diameter on the piece is obtained.

The positive stop H in Figure 17-10 is provided to make certain that pieces are formed to accurate diameters. The most practical method of setting the stop is to advance the form tool by means of the nut C until a sample approximately 0.008 inch under size is produced. With the cam roll on the full height of the cam, adjust the screw I until it holds the form tool back far enough to produce a sample piece of the correct size.

If the tool wears slightly, loosening the screw I slightly can correct the error. Be careful to set the tool to form parallel work. The two adjusting screws J in Figure 17-10, which shift the post with reference to its tongue, can correct the error. Before adjusting these screws, the nut G must be loosened slightly.

Then, the machine is ready to start. First, make a blank, and measure it carefully to see that all the tools have been set correctly.

Adjusting the Threading Tool

After the machine has been set to produce a suitable blank, it should be stopped just as the cam roll is positioned at the beginning of the thread lobe. Place the tap (or die) well back into the turret hole so that it will go only a short distance onto the piece.

If a solid die is used, make certain that the die is centered. If an opening die is used, the spindle pulleys can be run in the same direction. The slow speed can be used for threading, and the fast speed can be used for turning, drilling, and so on. It is also necessary to use a die closing arrangement when an opening die is used.

In starting the machine, make certain that the spindle is running in the correct direction and at the correct speed. Start the machine, and make certain that the spindle reverse trip dog has been set correctly to operate just as the cam roll drops off the high point of the thread lobe. With the tap, or die, placed a short distance outward, make a sample piece. Check the dimensions, and continue to adjust until the correct length of thread is obtained.

Setting the Deflector

Clamp a dog in position on the right-hand side of the carrier Y in Figure 17-6. On light pieces and work that can be machined quickly, the deflector must remain under the chute a greater portion of one revolution of the camshaft than on heavier pieces that drop instantly or on work that is machined slowly.

Setting the Automatic Stock

Adjust the spring in the tube until, with no stock in the spindle, its tension is great enough to throw the feed slide backward, thereby moving the roll on the feed operating lever from one side of the feed cam groove to the other. The machine will not operate if the tension is enough to overbalance the drag of the feeding fingers.

Measuring the Work

During operation of the machine, the pieces should be checked frequently. When working within fine limits, the work should be measured as each new bar is placed in the machine since a variation in stock can affect the cutting action of the tools.

Renewing Stock

When the feed is stopped automatically at the end of a bar, the chuck is left open. Before inserting a new bar, the turret locking pin should be withdrawn by the thumb latch. The turret should be turned to a point that is one-half the distance between stations so that the short piece of stock left in the chuck can be pushed out by the new bar. Make certain that all burrs are removed from the ends of the new bar. Turn the handwheel so that the chuck is open, and push the bar through until the end is just past the cutoff tool.

If the swing stop is being used, raise the stop by hand to allow the short end of the bar to drop out while inserting the new bar. Turn the turret back into position, and start the machine. The work should always be checked after insertion of a new bar.

Dial-Controlled Machines

The Cleveland Dialmatic machine has 112 spindle speeds, ranging from 24 rpm to 1820 rpm, and dial-controlled feeds. A two-speed motor supplies power for the spindle and provides two automatic spindle speeds. Two additional speeds are made available by an automatically shifted disc-type friction clutch. This makes a total of four automatically changed spindle speeds (both forward and reverse) that are available for each set of change gears.

Tool feeds are dial-controlled (Figure 17-14). An electric feed drive provides separate adjustable feeds for all turret positions. Feeds can be set quickly in these machines as follows:

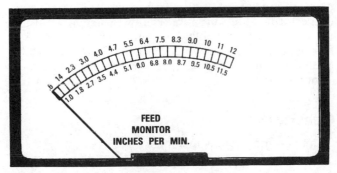

Figure 17-14   The monitor permits accurate feed settings in inches **per minute.** *(Courtesy Cleveland Automatic Machine Co.)*

- Cam changes are not necessary to adjust feeds.
- Spindle speeds and feeds can be transposed into dial setting by means of conveniently located charts.
- A turn of a knob sets a feed.
- Feeds in inches per minute are accurately indicated on the monitor.

- Ten different feeds are possible in a single cycle of the machine.
- Feeds can be adjusted while the tools are cutting.
- All feeds are infinitely variable.
- A lock on the cover can protect dial settings after setup.

Summary

The screw machine was originally designed to make small screws and studs. The flexibility of the machine has adapted it to a large variety of work.

A screw machine is a type of turret lathe designed to perform a series of operations at one time. The turret is provided with several tools, brought into a working position to perform various operations as the work is being turned.

Drilling, boring, reaming, forming, and facing are just a few examples of the operations that can be performed on a screw machine. The various types of screw machines can be classified as to the type of operation (such as plain or hand-operated, semiautomatic or automatic) and the type of spindle (such as single or multispindle).

Many different kinds of tools are used because of the wide variety of work performed on automatic screw machines. Turning tools, cutting tools, threading tools, knurling, forming, and cutoff tools are most generally used. Swing tools, fixed and adjustable guides for operating swing tools, backrests for turrets, and spindle brakes are other equipment that can be used on automatic screw machines.

Review Questions

1. What is an automatic screw machine?
2. How does the screw machine differ from the lathe?
3. What are the two types of screw machines, i.e., how are they classified?
4. What operations can be performed by the screw machine?
5. Describe what an indexing turret can do.
6. Standard cams control the forward and return motions of the _____.
7. Cutting lubricant must be used frequently to avoid excessive heat and to prevent ruining either the work or the _____.

8. List some of the tools used on the automatic screw machine.

9. What is the purpose of a knurling tool? What type of work does it do?

10. The _____ tool is used for taper turning and for some forms of turning.

11. What is the spindle speed for forward speed on the Brown & Sharpe No. 2 automatic screw machine?

12. Indexing the turret is performed by a simple _____ movement.

13. How do you change from a double to a single index?

14. The _____ tool is the one that governs the distance from the chuck more often than the others.

15. How do you adjust the turret to the correct distance from the chuck?

16. How do you set the stock for length?

17. How do you adjust the feed slide for length of stock?

18. How do you adjust the form tool?

19. How do you adjust the threading tool?

20. What is the range of spindle speeds for the Cleveland Dialmatic machine?

Chapter 18

Automated Machine Tools

Automation is sometimes confused with the use of computers in industry. The term *automation* is actually closer in meaning to the control of machine tools, which is only a single use for computers in industry. Computers are an aid in a wide variety of industries in the areas of inventory control, process control, and numerical control of machine tools.

The computer is used in effective inventory-control programs to provide management with the facts on which decisions can be based. It is used to forecast high-demand items, to adjust production to market, and to parallel more closely other trends with production schedules.

The computer is used in process control in industry or manufacturing to monitor by means of sensors, which measure whatever is going on. These measures are fed into a computer for control of the steps in the process. Signals from mechanical and electrical instructions, which are used to point out certain conditions and timing in the process or product, are used to control every industrial process. Normally, the operating personnel interpret these signals and transform them into control actions, but the input elements of the computer are designed to accept these same control signals so that the computer can interpret them and translate them into effective control actions. For example, industrial transducers that change pressure into voltage are used to convert these input signals into suitable voltages. The selected input groups are channeled to the computer logic in the correct sequence.

Control outputs from the computer can be converted to the required control actions, such as setting switches or turning valves. Thus, the computer actually controls the process, its timing, and its operation.

Numerical control for a system of automation uses a punched tape in which the holes represent coded instructions for the machine tool (Figure 18-1). These coded instructions tell the machine exactly where to drill a hole, and they give the exact size of the hole. This results in an automatic operation of the machine tool (Figure 18-2). This chapter is concerned chiefly with numerical control.

Figure 18-1 Programming and preparing tape is simple and time saving. It involves deciding on the number of workpieces per load and their location on the subplate, preparing the tool and program sheets from the piece part drawing with coordinate dimensions, and punching tape using conventional tape preparation equipment. *(Courtesy Moog Hydro-Point)*

Figure 18-2 A row of numerically controlled machining centers with 24-station random-selection automatic tool changers. *(Courtesy Moog Hydro-Point)*

Basic Principles of Numerical Control

One important aspect of computers and computer-like techniques is the numerical or digital control of machine tools. Numerical control is concerned with how the machine director or controller responds to commands expressed by numbers. The unit causes tile controls of the machine (such as handwheels or cranks) to be moved by means of power, rather than moved manually by an operator (Figure 18-3).

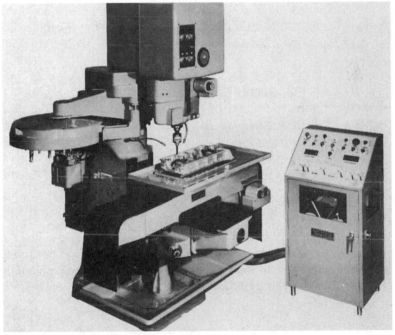

Figure 18-3 This numerically controlled machining center is composed of a milling machine and control unit. The control unit causes the controls of the milling machine to be moved by means of power rather than manually by an operator. *(Courtesy Moog Hydro-Point)*

Since automation is a method that is used to avoid physical labor, it is usually accomplished by designing machines that repetitively perform either single or multiple operations that would normally be performed by a machine operator. Numerical control is a method of industrial automation where the method used to avoid physical labor, instead of being a "fixed" repetitive method, can be changed at will by means of variable programming.

Numerical control is a type of programmed automation. Consider the common drill press, for example, as a power-driven tool. The vertical spindle of the drill press moves up and down in relation to the worktable, which normally is fixed in position. The workpiece to be drilled can be moved into position by maneuvering it on the worktable until the spot where the hole is to be drilled is directly below the twist drill.

Consider a 500-tooth gear and a lead screw pitch of one inch per revolution, so that the table moves one inch when the gear and shaft are rotated one revolution. If the gear is moved only 10 teeth, the slide will move only 0.02 inch; for a movement of one tooth, the slide will move 0.002 inch.

Stops are used on the gears. If there are 10 stops spaced equally around a gear with 1000 teeth, an amount of movement of the worktable (in decimal parts of an inch) can be selected. The first stop will let the gear rotate $\frac{1}{10}$ revolution, or 100 teeth. If a train of four gears is involved, each having 10 stops, and the gears are connected in a gear ratio of 10 to 1, the first gear and the interposers will control the table movement in 1-inch steps within a 10-inch measure. The second gear will control the table movement in 10 steps of 0.001 inch. It is possible, by selecting the proper stop on each of the four gears, to obtain stops that control movement to the precision of a four-digit number, such as 2.357 inches.

Switches can be set up manually, or a punched tape and a tape reader can be used to decode a set of holes for each digit. Thus, the slide table on the drill press will move the correct distance for the dimension as was programmed in the tape.

Several other systems can be used, two of which use electronics. One system uses voltages that are related to digits in the dimensions. The worktable has a transformer that produces a voltage proportional to its movement. The movement of the worktable stops when the two voltages are equal.

Another electronic system uses pulses. The dimension is given in terms of the number of electrical pulses. For example, if the dimension is 42.3715 inches, there are 423,715 pulses to be counted. In this system, the drive motor stops when the number of pulses counted equals the number set up in the predetermined counter.

Pulses for counting can be produced by several methods. One of the most accurate methods is the use of a finely graduated scale on a bar of glass connected to the worktable. In this system, these graduations are a series of fine parallel lines on the glass bar. Using a light source on one side and a photocell on the other, a pulse is produced for each line on the glass bar when the beam of light to the photocell is broken by the line. Line width and spacing between lines are equal. As many as 10,000 lines per inch are possible.

A different method uses a steel bar that has magnetic spots on the worktable. A magnet pickup coil is mounted near the magnetic spots on the bar. When these spots pass by the pickup coils, a voltage pulse is generated, and pulses are then counted in special circuits. The pulse number is set up into a predetermined or preset counter. There is a difference between tile pulse number and the number registered from the scale, but a closed control relay circuit supplies the power to an electric motor which, in turn, drives the gear and the lead screw, moving the table in the proper direction for the pulses to be counted. When the two pulse numbers are the same, this is recognized, and the motor stops at this point. As a result, the worktable stops in the proper position.

Many applications of numerically controlled manufacturing processes can be found. One example is reversing-tape rolling mills for aluminum and steel, where the ingot moves back and forth through a series of rollers that force the metal into thin bars and then into thin sheets. As the opening between the rollers must be changed and made smaller and smaller in this process, close control of the gage screws is required. These are turned until the bar is finally reduced to the desired thickness. The rate and amount of turning down varies with the material, temperature, and other characteristics. This action is controlled by a program using either a punched tape or a punched card.

Numerically controlled ultraprecision inspection and measuring machines are in use. The machine tool industry has made wide use of this principle and has found that it can be applied to many processes and operations including drilling machines, boring machines, milling machines, lathes, shapers, punch presses, flame cutters, and welders.

Preparation for Numerical Control

The Armour Institute of Chicago was one of the first to study the field of numerical control. They applied it to a lathe. The Massachusetts Institute of Technology applied numerical control to a milling machine in 1952. The operation used punched paper tape. Tape preparation equipment and a tape reader were used to place the numerical-control information into the machine tool director.

Several types of equipment are now available because of expansion of the employment of numerical control in the industrial field.

The first step in numerical control planning is to start with the raw data or engineering drawing of the part. The process planner must study the part, determine just how it is to be made, and plan to use the standard tools (such as drills, reamers, cutters, and holding fixtures) or provide for special tooling if it is required. Special tools, which are usually holding fixtures, are sometimes required when the part cannot be mounted in the usual manner on the worktable of the machine tool by means of the fixture in stock.

After recording all the necessary information on a planning sheet, the process planner must check the sheet for accuracy to make certain that the processing steps are in the correct sequence and will work. The next step, after completion of the planning sheet, is to have it typed on a machine that produces a punched paper tape or control tape (Figure 18-4). The control tape is checked carefully for accuracy of interpretation from the source copy and for machine accuracy to make sure that it will work

Figure 18-4 Conventional tape preparation unit. This unit is usually located away from the machining area. *(Courtesy Moog Hydro-Point)*

The control tape can be retained as a master control tape and a duplicate tape made for use in the shop. The tape is placed in the tape reader on the machine director to run the machine for production of the required parts (Figure 18-5). Usually, the tape is used to run the machine through all the steps in the numerical-control operation without any work on the worktable to make sure that a step has not been overlooked, or incorrectly programmed, and that production is possible. To illustrate preparation for the numerical-control operation, the drawing of the casting (Figure 18-6) can be used as the starting point for programming. All required dimensions of the part are taken from the drawing and placed on the process planning sheet, along with the other information.

In addition to the x- and y-coordinates (Figure 18-7), a considerable amount of information is required. The tool group and number and the auxiliary functions to be performed are all required, as

Figure 18-5 Control unit for a machining center. In addition to providing the capability of making the initial tape, this unit also enables the operator to make changes or corrections easily and quickly right at the machine, thus eliminating the need to send the original tape back to the programming department for editing. This unique advantage can reduce setup time by as much as 40 percent. *(Courtesy Moog Hydro-Point)*

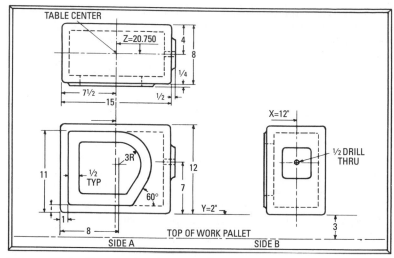

Figure 18-6   Engineering drawing of a casting that is to be produced by numerical-control operations.

well as the x- and y-coordinates for each step. Programming can be one-quadrant or four-quadrant, as shown in Figure 18-7.

The final programs or numerical-control tapes vary in length, depending on the number of operations, but most tapes range from 19 to 60 inches in length. Some tapes are much longer.

The Electric Industries Association on Numerical Controls has a standard one-inch tape with as many as eight holes for coding. Several other organizations, such as the National Machine Tool Builders Association and the Aerospace Industries Association, have agreed to adopt these same standards for their organizations. Typical coding, in which the binary-coded decimal system is used, is shown in Table 18-1.

The unpunched paper tape has an overall width of 1.000 inch ±0.003 and a thickness of 0.004 inch ±0.003 inch and can be used for recording six, seven, or eight levels of information across the tape.

A standard format of the program in a punched tape permits the tape prepared for use on one machine to be used on other machines. A tape prepared for a given two-axis numerically controlled machine should work on all two-axis numerically controlled machines.

Various programming formats are possible. Figure 18-8 compares four methods of programming that can be used for a

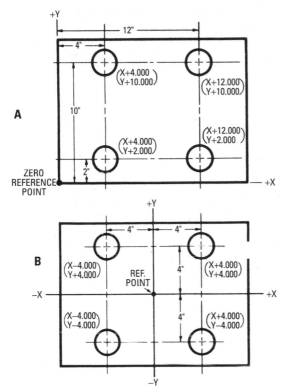

Figure 18-7 Programming a part for production by numerical-control methods: (A) one-quadrant programming; (B) four-quadrant programming.

numerical-control tape prepared for the same item. The four systems are *fixed sequential, block address, word address,* and *tab sequential.* Also, a format is shown for a universal tape, which can be used for any one of the other systems.

Electronic Control of Machine Tools

Automation means control of manufacturing. In electronic control of machine tools, a recording is made of the various movements required of tile machine tools for a specific series of operations that are required to produce a machine part. The recording, which is translated from a drawing of the part, is recorded by means of either a magnetic tape or a punched paper tape.

Table 18-1 Binary-Coded Decimal System
for Numerical Control

Standard Track Numbers									Digit or Letter Codes
8	7	6	5	4	.	3	2	1	
		6							0
								1	1
							2		2
			5		.		2	1	3
					.	3			4
			5		.	3		1	5
			5		.	3	2		6
					.	3	2	1	7
				4	.				8
			5	4	.			1	9
	7	6			.			1	a
	7	6			.		2		b
	7	6	5		.		2	1	c
	7	6			.	3			d
	7	6	5		.	3		1	e
	7	6	5		.	3	2		f
	7	6			.	3	2	1	g
	7	6		4	.				h
	7	6	5	4	.			1	i
	7		5		.			1	j
	7		5		.		2		k
	7				.		2	1	l
	7		5		.	3			m
	7				.	3		1	n
	7				.	3	2		o
	7		5		.	3	2	1	P
	7		5	4	.				q
	7			4	.			1	r
		6	5		.		2		s
		6			.		2	1	t
		6	5		.	3			u
		6			.	3		1	v
		6			.	3	2		w
		6	5		.	3	2	1	x

Table 18-1   (continued)

Standard Track Numbers	Digit or Letter Codes
6 5 4 .	y
6 4 . 1	z
7 6 4 . 2 1	. (period)
6 5 4 . 2 1	, (comma)
7 6 5 .	+ (plus)
7 .	− (minus)
6 5 . 1	V (check)
7 6 5 4 . 3 2 1	Delete
8 .	End of Block
4 . 2 1	End of Record
5 .	Space
6 4 . 2	Back Space
6 5 4 . 3 2	Tab
7 6 5 4 . 3	Upper Case
7 6 5 4 . 2	Lower Case

Figure 18-9 shows typical modern units used for numerical control of machine tools. Four steps involved in their use are development of the blueprint, planning, preparation of a control tape, and running the tape on the machine.

The numerical-control systems are basically incremental point-to-point positioning systems of the on–off tape using binary-coded decimals. As many as five axes can be controlled either simultaneously or sequentially with the basic system. System resolution can be either 0.001 inch or 0.0001 inch, with only nominal differences in the electronics.

The block diagram illustrated in detail in Figure 18-10 shows one axis of the control. All the control elements except the tape reader, distributor, and sequence control are repeated for the other axes. This is typical of a machine tool system that machines automatically and performs single-point boring operations from either end with infinite center locations and variable bore diameters, within the range of the machine. Tape-control positioning equipment is used with the capability of controlling the size of bores during the machining cycle. Table 18-2 provides a summary of the numerical-control specifications that are possible in the system shown in Figure 18-10.

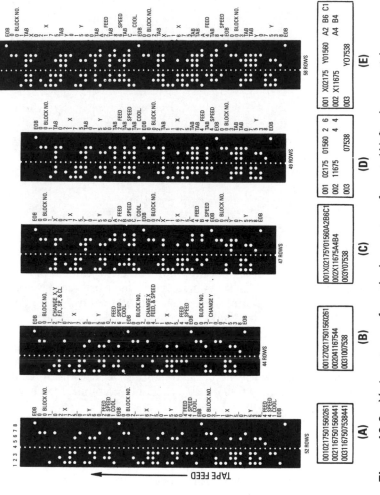

Figure 18-8 Various types of punched paper-tape format: (A) fixed sequential, (B) block address, (C) word address, (D) tab sequential, and (E) universal.

416

Figure 18-9   Automatic numerical-control system.
(Courtesy Excello Corporation)

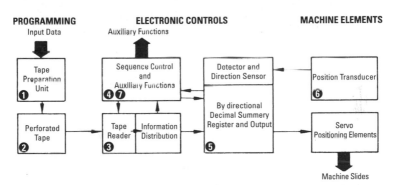

Machine Slides

Figure 18-10   Numerical-control system block diagram. *(Courtesy Moog Hydro-Point)*

Three traditional axes are used: longitudinal (x), horizontal (y), and vertical (z). A fourth axis of control is necessary for boring holes of varying sizes. The fourth axis is the tool or radius axis (r) that permits control of the programming of an infinite number of boring diameters.

The addition of the tool or radius axis required the development of a tool control mechanism capable of accepting numerical commands.

Table 18-2   Summary of Numerical-Control Characteristics

	GENERAL	TYPICAL MODELS*			
	Per Requirements	M23	M33	M44	M54
Number of Axes2		2	3	4	5
Travel	Virtually unlimited in all axes	99.999″	99.999″	X 99.9999″ Y 99.9999″ Z 99.9999″ R 0.9999″	X 99.9999″ Y 99.9999″ Z 99.9999″ R 0.9999″
Resolution	00001 in. (all axes if desired)	0.001 in. all axes	0.001 in. all axes	0.0001 in. all axes	X 0.001 in. Y, Z, R 0.0001 in.
Positioning Speed	Unlimited depending upon resolution only	180 in./min.	180 in./min. Z - 210 in./ min.	100 in./min.	120 in./min.
Tape	1 -in.-wide, 8-channel AI-Mylar laminate	Same	Same	Same	Same
Tape Reading Speed	100 lines/sec.	Same	Same	Same	Same
Auxiliary Functions	Per requirements	11	11	20	20
Feed Rates	Large number available in any or all axes	—	—	0.5 to 20 in./min. in 31 steps	0.5 to 10 in./min. in 16 steps
Operation Modes	Automatic, semiautomatic, manual	Same	Same	Same	Same
Voltage Requirement	115AC, single phase	Same	Same	Same	Same

Courtesy Ex-Cell-O Co.

In this device, an electric servomotor is used to position the boring bar to change boring size in response to punched-tape signals.

There are three modes of operation:

- *Dial-in,* or *manual, mode*—This is used for very short runs. The decade counter dials are set to the required dimensions. The axis direction switches are set, and the "start" button pushed. The dimensions are displayed decimally. When the "manual-start" button is pressed, the table moves to the dimensions immediately.

- *Semiautomatic*, or *block*, *mode*—This is used for tape verification and general checking. The tape is inserted in the tape reader and the "read" button pushed. The photoelectric reader reads the tape at 100 times per second, and the dimensions are displayed decimally on the decade counter tubes of the summing register, where they can be compared with the blueprint or planning sheet dimensions. In manual or block mode, the spindle (if applicable) can then be actuated.

- *Automatic mode*—This is normally used to operate the machine tool automatically through programmed functions. The tape is inserted in the reader and the "read" button pressed. The photoelectric reader reads one block or a complete machine command.

At the end of the reading phase, the side axes move to position simultaneously, and the spindle is then activated. The operator can always check the accuracy of the system by watching the position display.

The system block diagram consists of three parts: programming, electronic controls, and the elements of the machine tool itself. Any number of axes can be controlled simultaneously or in a specified sequence. Each axis of control is complete in a chassis that slides into the main cabinet.

The output from this system can control electric drive motors or hydraulic servo valves that control hydraulic motors or pistons. Feed-rate control can be added for applications that require various controlled feed rates.

Tape Preparation

In reference to Figure 18-10, a tape preparation unit (1) is used to prepare a punched paper tape (2) (refer to Figure 18-4). The input unit is a tape reader (3) that operates at a speed of 100 lines per second. No machine time is lost for data input. Tape information is distributed to auxiliary functions and sequence control (4) and to the summing register's computer (5) by electronic switching.

The data storage medium in this instance (Figure 18-11) is standard 1-inch, eight-channel paper, *Mylar*, or *Al-Mylar* tape. The tape code is EIA-approved binary-coded decimal, and the tape format is tab sequential. The tab markers (code holes in the tracks number 2, 3, 4, 5, and 6) are punched to separate information code words, such as *x-axis* or *y-axis*.

The end-of-block code can be punched at any line of the tape denoting that no more information is forthcoming for that machine

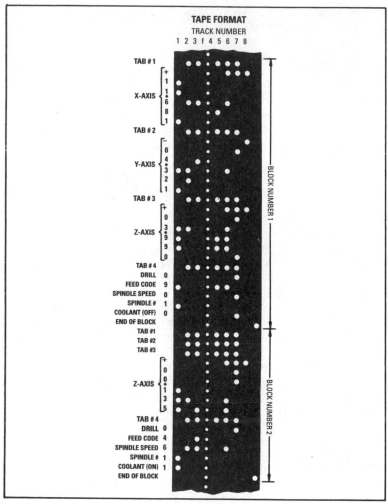

Figure 18-11 The tab-sequential tape format used with the machine.

command. Only the tab marker need be punched when no movement is required of an axis. It is unnecessary to add unused axis dimensions. Thus, in the operation shown in block No. 2, only the tab markers are punched for X and Y.

Control

The tape reader reads one complete operation (distance to be moved and direction of motion for each axis) into the bidirectional register (5) in Figure 18-10 whose output controls drive the axes simultaneously in the command direction at rapid-traverse speed. The position transducers (6) in Figure 18-10 convert motion into numbers for feedback signals indicating direction and distance of motion. The register also counts the distance moved, compares it to what it should be, and makes corrections. When all axes of the tool are in position, an "In-Position" lamp is turned on. An electric signal is then available for the tool feed. The control will read the next operation from the tape when it receives a "tool-feed cycle finished" signal for an automatic cycling machine, or a push-button signal for the operator of a machine that has a tool-feed cycle.

The sequence control (4 and 7 in Figure 18-10) divides the system into two basic operating phases—the read phase and the machine phase. When the end-of-block signal appears after a block of tape has been read, the sequence control stops the tape reader and switches from the read phase to the machine phase. The machine phase "start" signal allows axis movement. When the machine element reaches the programmed position, a move relay is de-energized, and a "coincidence" signal appears at the sequence control. When all axes have reached coincidence, the sequence control switches from machine phase back to read phase, and the cycle is repeated for every block programmed on the tape. An end-of-tape signal prevents the sequence control from entering the read phase again.

At the beginning of the read phase, a tape-start signal is sent to the summing register to clear any spurious counts and to inhibit the borrow-and-carry function, because decimal information is entered into each decade in parallel. At the end of the read phase, a tape-stop signal is transmitted to the summing register to free the counter for the machine phase.

During the read phase, each of the several decades of the summing register receives decimal information in parallel from the electronic distributor. This programmed dimension represents the incremental distance to the next slide position and is momentarily displayed in the decimal counting tubes prior to the machine-start phase. During the machine phase, pulses are received by the register from the position transducer. The pulses are serially fed into the least significant decade of the bidirectional counter. These pulses return the counter to zero.

Transducers

The sensing elements of the position transducer are two magnetic heads that sense the rotation of the shaft. The unit can be applied either to precision ball-loaded lead screws or to precision-measuring racks with pinions.

During the machine phase, the transducers feed back digital pulses, indicating distance and direction of movement. These pulses are fed serially into the least-significant decade of the bidirectional counter. The counter counts down until the slide is at a predetermined distance from the stopping point. At this time, the slide velocity is precisely set to the controlled approach rate. At the stopping point, the last pulse to the counters stops the slide drive motors, activates the slide lock, and triggers the next function if applicable.

The number of auxiliary functions that can be controlled is almost unlimited. Examples of these functions are spindle or tool selection spindle speeds, coolant on–off, and feed-rate control. Feed rate is generated as programmed from the tape for the desired feed-rate characteristics in applications where it is necessary.

Machine sequencing and output controls are mechanized with relays to meet the requirements of a particular machine and its positioning elements.

Figures 18-12 through 18-19 show a number of close-up views of machining operations performed on the numerically controlled machining center shown in Figure 18-2 and Figure 18-3.

Figure 18-12 Peck drilling. A pecking action is used to facilitate deep hole drilling. The cutting time between pecks is preset on a timer, eliminating the necessity to program each peck depth. At each peck, the tool is withdrawn (at 300 ipm) to clear the chips and returns (at 300 ipm) to 0.015 inch above the previous cut. This ensures maximum tool life and gives accurate deep holes, as small as 0.014 inch in diameter, without tool breakage. *(Courtesy Moog Hydro-Point)*

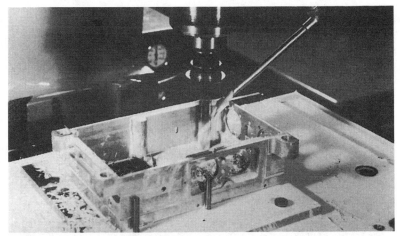

Figure 18-13   Pocket milling can be carried out with guaranteed accuracy in all three axes. The backlash-free positioning system permits milling in both directions on each axis, and climb milling can be programmed to produce the best possible surface finish. *(Courtesy Moog Hydro-Point)*

Figure 18-14   Boring. A super-precision quill fit coupled with guaranteed spindle runout less than 0.0002 in TIR and positive tool orientation ensures close-tolerance bored holes. Hole size and geometry can be held within 0.0002 inch under production conditions. This "canned cycle" allows tool withdrawal at programmed feed rate, thus eliminating taper and tool marks. *(Courtesy Moog Hydro-Point)*

Figure 18-15 Fly cutting. Excellent finishes can be obtained using up to 5-inch-diameter fly cutters. Up to 3-inch diameter can be loaded in adjacent stations. *(Courtesy Moog Hydro-Point)*

Figure 18-16 Heavy milling cut. Machines will remove 2.3 cubic inches per minute of mild steel. *(Courtesy Moog Hydro-Point)*

Figure 18-17 Coordinated tapping provides a fast tapping cycle, where spindle speeds and feeds are coordinated under tape control, to match the pitch of the thread. A compensating tension/compression tap holder is used for the operation. *(Courtesy Moog Hydro-Point)*

Figure 18-18 Drill cycle. The standard drill cycle can be used to perform boring and reaming operations where rapid retraction of the tool is acceptable. *(Courtesy Moog Hydro-Point)*

Figure 18-19 The lead-screw tapping operation will produce class 3 threads with very accurate depth control. A precision-ground lead screw is hydraulically traced, and it controls the feed rate of the tap in relation to the pitch and spindle speed. The only load applied to the tap is the torque required for cutting. *(Courtesy Moog Hydro-Point)*

Summary

Automation is a method that is used to avoid physical labor and is accomplished by designing machines that repetitively perform either single or multiple operations that would normally be performed by a machine operator. Numerical control, instead of a "fixed" repetitive method, can be changed at will by means of variable programming.

Various automation systems can be used. Switches, or punched tape and a tape reader, can be used to decode a set of holes for each digit. Other electronic systems utilize voltage pulses. Pulses for counting can be produced by several methods. One of the most accurate methods is the use of a finely graduated scale on a bar of glass connected to the worktable.

Review Questions

1. What is the meaning of the term *automation*?

2. Numerical control for a system of automation uses a punched _____ in which the holes represent coded instructions for the machine tool.

3. One important aspect of computers and computer-like techniques is the numerical _____ control of machine tools.

4. What is the function of the control unit?

5. Numerical control is a type of _____ automation. What is the more modern way?

6. Who was one of the first to study the field of numerical control?

7. The final programs or numerical-control tapes vary in length, depending on the number of operations, but most tapes range from 19 to _____ inches in length.

8. What are the dimensions (width and thickness) of the tape used to control machines?

9. What are the four methods of programming?

10. What are the four steps involved in the use of modernized tape-controlled machines?

11. What is a transducer?

12. What is peck drilling?

13. Describe pocket milling.

14. What is fly cutting?

15. What is a lead screw?

Chapter 19

Computerized Machining

Most newer machine tools are now computer controlled. Computer Numerical Control (CNC) means that a computer controls the movements of a machine by computer programs that are made up of letters, symbols, and numbers. Milling machines, drill presses, grinders, turret punches, welding machines and lathes are some of the more common types of CNC machine tools. Programs for CNC machines contain step-by-step instructions to tell the machine every movement of the tool or workpiece, the cutting speeds and the feeds of each machining operation, the start and end of each machining operation, and the start and end of coolant flow. Once a program has been written for a CNC machine, parts can be produced automatically by the machine each time the program is run (Figure 19-1).

Skilled machinists are no longer needed to run the CNC machines. The machining is done automatically according to the information that is recorded on the program. However, skilled machinists are needed to write the programs to make sure the work is done correctly. The computer has the ability, when properly utilized, to improve the quality of the finished product inasmuch as it can do the same operation over and over again without becoming tired or bored.

CNC has several advantages over the manually operated and controlled machines. CNC is more accurate and reliable. Fewer parts are spoiled because of human error. CNC programs can be stored on floppy disks, magnetic tape, or any of the other available memory devices. Programs have the capability of being returned quickly to the machine for reuse. This eliminates the need to replan the job. Programs can be prepared faster than jigs can be designed and built for manually operated machine tools. This saves the company time and money.

CNC also has a few disadvantages. CNC machines are more complex than manually operated machines and are, therefore, more expensive to purchase. They cost more to maintain since they are more complex. Highly skilled maintenance personnel are required to keep the machines operational. Programmers for CNC machines require a high level of training and are hard to find. It is rare for a machinist with a few years of experience to have a

```
++INPUT AND TOOLING
PART            TAPERED SLEEVE              210725695   30
COMMENT         FOR W.T.S.C. BY D.H.G.
CLASS           02
COMMENT         CAST IRON CLASS    30
MACHINE         SC28
GRIPDIAM        12.25
SPIN-LOC        10.
LOC-INT         .875
PRTLENTH        5.875
FRNTFACE        5.875 .375        .005      125
EXTERNAL
CHAMFER         *     .0620   -0.       -0.         -0-0    45.0000
DIAM            *    6.5010   6.4490   8.5000      125-0
THREAD              1.3750    .1250   -0           -0-0
NECK            *     .1250    .2500   -0           -0-0
FACE            *    1.5000   -0.       1.000      125-0
TAPER           1    -0.       2.0620  -0.          -0-0    15.0000
DIAM            *    8.0000    .0020   8.5000       63-0
RADIUS          *    -0.      -0.       -0.         -0-0     .2500
FACE            *    4.5000    .0050    .3750      125-0
RADIUS          *    -0.      -0.       -0.         -0-0     .1250
DIAM            *   11.7500   -0.     12.2500      250-0
FACE            *    4.8750   -0.       .7500      250-0
CUTSHARP
DIAM            *   12.2500   -0.     12.2500       -0-0
INTERNAL
RADIUS          *    -0.      -0.       -0.         -0-0     .3120
DIAM            *    4.5010   4.4990   3.5000       63-0
RADIUS          *    -0.      -0.       -0.         -0-0     .0620
FACE            *    3.0000   -0.      3.0000      125-0
CHAMFER         *     .0150   -0.       -0.         -0-0    30.0000
DIAM            *    4.0000   -0.      3.5000      250-0
FACE            *    5.8750   -0.       -0.         -0-0
CUT-PAST
ENDPART
```

WARNER & SWASEY CUTS VERSION 2 MODIFICATION 2   09/01/02   08.41.46

STATION	TOOL	OP NAME		HOLDER	INSERT	RADIUS	ZOFSET	XOFSET
1	10	TURN	6503	12L-HEAD	TPG544	.0625	15.3125	-3.9833
2	80	ETRD	6506	L-BLK	NT-4R	.0100	15.8720	-3.6450
3	20	FACE	6503	12L-HEAD	TPG544	.0625	15.2667	-4.1875
4	43	RBOR	2835	3.50DBAR	TPG433	.0469	15.0781	-.7938
	53	FBOR	2835	3.50DBAR	CPG422	.0313	15.0906	2.5906
5	30	FINI	6503	12L-HEAD	CPG632	.0313	15.3406	-4.1594
6	60	EGRV	6506	L-BLK1/4	NG-48R	.0100	15.9900	-3.6350

START STATION NO. 1, 16.0 INCHES FROM THE STOPS, RADIAL SCALE 0.0

CAUTION - CHECK UNUSED STATIONS FOR INDEXING CLEARENCE.

TOTAL TIME 18.2 MINUTES
LAST BLOCK NUMBER IS 275
TAPE LENGTH = 53 FEET 10 INCHES

Figure 19-1   Program for a CNC machine.

computer background. Therefore, it takes additional time and money to train personnel for specific CNC equipment.

Numerical Controls

The Allen-Bradley Company makes a low-cost, full-featured control specifically designed for mills and machining centers with up to six axes. It uses compact construction for ease of installation on new or retrofit machines, as well as a color graphics display. Simplified prompting and advanced software features allow for highly productive CNC control of machine operations.

The Bandit IV CNC (Figure 19-2) has a 12-inch cathode ray tube (CRT) graphic display that assists in design, debugging, setup, and production. The operator is provided with a visual record of both the part and the machine movements. Dry runs can be completed and the outcome checked against the graphic display. This makes it easy to troubleshoot and make necessary corrections. A special window or zoom feature allows the operator to zoom in on a specific section of the programmed plot to examine it in more detail (Figure 19-3).

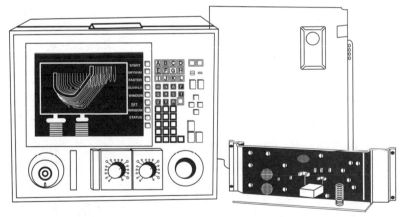

Figure 19-2   Bandit IV is a full-featured control specially designed for mills and machining centers with up to six axes. *(Courtesy of Allen-Bradley Co.)*

Some of the features include cavity programming, part program memory expansion, programmable back divide, polar coordinate programming, auto cycles, helical contouring, plane switch capability, extensive editing capability, digitizing, rotary axis capability with rollover, multiaxis simultaneous jog moves with handwheel, and look-ahead cutter compensation.

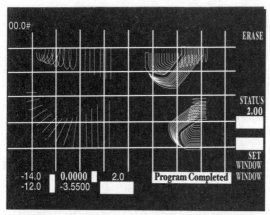

Figure 19-3 CRT readout for the Bandit IV. *(Courtesy of Allen-Bradley Co.)*

This unit can be integrated into new or existing machines quickly and easily. It uses a new type of connector to allow controls to be connected without the use of special tools.

Computer-Operated Machine Tools

Computer control of machine tools became feasible around the mid-1960s with the introduction of the integrated circuit. Integrated circuits (also known as *chips*) are made for computer use and can be adapted to operate machines such as lathes, milling machines, and shapers. The low-cost control circuitry has made it feasible to produce everyday products on the machine without human supervision at all times.

Computers in the late 1960s were still quite large and expensive compared with today's standards. Because of this, *direct numerical control* (DNC) became the first type of computer control program. A DNC system is made up of one large computer that can operate many machine tools on a time-shared basis. Following are some of the advantages of a DNC system over a conventional NC system:

- The tape reader and punch tape are eliminated from the system.
- Programs can be used to run several machines.
- Programs can be easily edited without having to punch an entirely new tape.

In the early 1970s, computers became a lot less expensive and more compact. They were also rugged enough to be subjected to harsh environments that may be encountered on a shop floor. These advances in computer technology led to the introduction of the computer numerical control (CNC) system. These systems are *characterized* by having a computer built into the machine tool. Each machine tool, therefore, has its own computer. The advantage of this system over a DNC system is that it is a lot less expensive and only one machine tool goes down when the computer fails.

Today, CNC machine tools can be linked to a mainframe computer. This combination (commonly called *distributed numerical control*) is an updated version of direct numerical control. A distributed numerical control system has all the advantages of the earlier direct numerical control system along with the ability to communicate with the CNC machine tools on the shop floor. This allows the mainframe computer to keep track of such things as the number of parts that are being produced by each machine and the number of times each machine is down for maintenance. The combination of DNC and CNC also alleviates the problems of machine tool downtime when the mainframe computer fails. This is because the computer built into a CNC machine tool can be operated in a stand-alone fashion.

CNC Components and Control System

The two main components of a CNC machine are the machine tool and the *machine control unit* (MCU). The MCU is the computer that operates the machine tool. There are many manufacturers of these control units, so program codes may vary somewhat between machines.

Figure 19-4 shows three types of control systems for NC and CNC machines. The least-expensive type is called a *point-to-point* (PTP) *control system*. A PTP control system can only move a tool in a straight line and is limited to hole operations (drilling, boring,

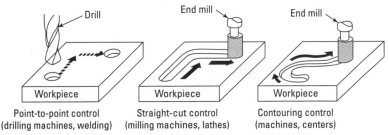

Point-to-point control
(drilling machines, welding)

Straight-cut control
(milling machines, lathes)

Contouring control
(machines, centers)

Figure 19-4   Point-to-point control system.

reaming, etc.) and spot welding. The tool path between points is not controllable in a PTP system.

A *straight-cut control system* is capable of moving the cutting tools in a straight line between points. However, only one axis drive motor can move at one time, so arcs and angles are not possible.

The *contouring control system* is the most expensive and flexible of the three systems. With the recent advances in electronics, the price difference between these systems has narrowed so much that every CNC machine is now manufactured with a contouring control system. A contouring system is capable of doing what the other systems can do while also being able to cut arcs and angles.

Positioning Formats

A CNC program is written using the Cartesian coordinate system. The CNC tool can be moved inside the rectangular coordinates in either two or three dimensions (Figure 19-5). Lathes are

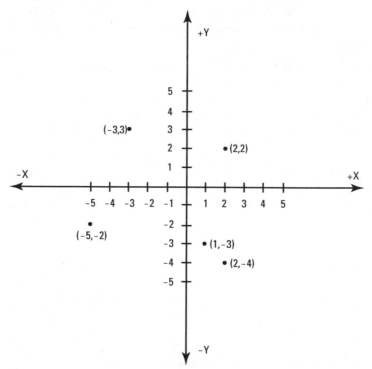

Figure 19-5 Cartesian coordinates.

programmed two-dimensionally using only the x- and y-axes, while milling machines are programmed three-dimensionally using the x-, y-, and z-axes (Figure 19-6).

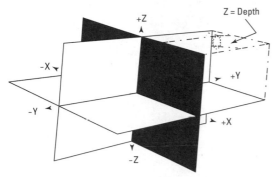

Figure 19-6 Three-dimensional NC uses the z-coordinate. Most machines have the z measurement begin when the tool spindle is fully retracted. Positive values are above the point of origin and minus values are below.

The z-axis is always the axis that determines the depth of cut. The x-axis is parallel to the longest length of the machine table, and the y-axis is parallel to the shortest length of the machine table. CNC programs can be written in one quadrant or four quadrants just like NC programs.

CNC machines can be positioned in two ways with respect to the Cartesian coordinate system. They can be positioned in an absolute format or an incremental format. In the *absolute format*, all machine locations are taken from starting point (0,0), which is referred to as the *home position* (Figure 19-7). There are no assigned or home coordinates in the *incremental format*. The cutter tool bit moves in spaces (increments) relative to the coordinates previously listed (Figure 19-8).

The absolute format is typically used for the main body of a CNC program. This allows the cutting tool to be given a home position to start and end its job. The incremental format is used in the middle of a program when a repetitive cycle (commonly referred to as a *canned cycle*) is needed (Figure 19-9). A canned cycle is a cutting pattern that can be programmed to repeat as many times as desired. It can reduce the size of a program tremendously, depending on the number of machining operations involved. The size of a program is reduced by a canned cycle

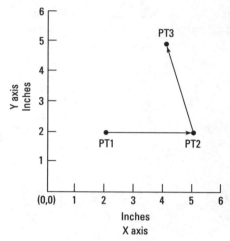

Figure 19-7 Absolute system. All points are relative to (0,0).

Coordinate Positions

Point	X value	Y value
PT1	2	2
PT2	5	2
PT3	4	5

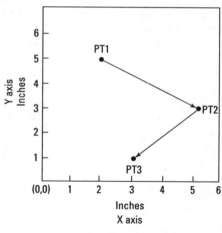

Coordinate positions

Point	X value	Y value
PT1	2	5
PT2	3	-2
PT3	-2	-2

Figure 19-8 Incremental system. All points are in reference to preceding point.

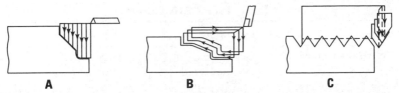

Figure 19-9 (A) Stock removal. (B) Pattern repeating. (C) Thread cutting.

because a programmer does not have to give the fixed coordinates of each move of the cutting tool. A programmer must only give the distance from each previous move of the cutter for one cycle. After that, a programmer merely has to list the number of times the cycle must be repeated. A programmer can also call up the canned cycle and use it again to do a similar machining operation in another part of the workpiece.

Advantages of CNC over NC

Conventional NC machines that are not connected to a computer do not have the memory capabilities to run a canned cycle. Every move of the cutting tools would have to be listed coordinate by coordinate. NC machines also lack the power of a computer's memory to calculate arcs and curves instantaneously. CNC machines have the added feature of graphic simulation. This feature allows the programmer to watch a visual display of how the CNC machine will machine the part according to the program that was just created. Any errors in the program will be displayed so that a programmer can debug a program without wasting material or tool bits.

CNC Programming

Because of the added feature of a computer's memory, CNC programming is a lot easier and simpler than conventional NC programming. Most CNC machines can be programmed using a standardized list of codes that were developed by the Electronics Industries Association (EIA), as shown in Table 19-1. These codes include preparatory functions or G codes. Some of these functions include cutting an arc in a clockwise direction (G02-circular interpolation) and telling the machine that the dimensions will be given in millimeters (G71-metric programming).

Table 19-1 EIA Codes

Code	Description
Preparatory Functions	
G00	Denotes rapid traverse for point-to-point positioning
G01	Linear interpolation
G02	Circular interpolation clockwise
G03	Circular interpolation counterclockwise
G04	Dwell
G05–07	Unassigned
G08	Acceleration at a smooth rate
G09	Deceleration at a smooth rate
G10–16	Unassigned
G13–16	Axis selection codes
G17	XY plane selection
G18	ZX plane selection
G19	YZ plane selection
G20–32	Unassigned
G33	Thread cutting, constant lead
G34	Thread cutting, increasing lead
G35	Thread cutting, decreasing lead
G36–39	Unassigned
G40	Cutter diameter compensation cancel
G41	Cutter diameter compensation left
G42	Cutter diameter compensation right
G43	Cutter diameter compensation inside corner (used to adjust for differences in programmed and actual cutter size)
G44	Cutter diameter compensation outside corner (used to adjust for differences in programmed and actual cutter size)
G45–49	Unassigned
G50–59	Used with adaptive controls
G60–69	Unassigned
G70	Inch programming
G71	Metric programming
G72	Three-dimensional circular interpolation clockwise
G73	Three-dimensional circular interpolation counterclockwise

Table 19-1 *(continued)*

Code	Description
G74	Multiquadrant circular interpolation cancel
G75	Multiquadrant circular interpolation
G76–79	Unassigned
G80	Cycle cancel
G81	Drill cycle
G82	Drill cycle with dwell
G83	Intermittent or deep hole drilling cycle
G84	Tapping cycle
G85–89	Boring cycles
G90	Absolute positioning
G91	Incremental positioning
G92	Register preload code
G93	Inverse time feed rate
G94	Inches (millimeters) per minute feed rate
G95	Inches (millimeters) per revolution feed rate
G96	Unassigned
G97	Revolutions per minute spindle speed
G98–99	Unassigned
Miscellaneous Functions	
M00	Program stop
M01	Optional (planned) stop
M02	End of program
M03	Spindle on clockwise
M04	Spindle on counterclockwise
M05	Spindle off
M06	Tool change
M07	Coolant on (flood)
M08	Coolant on (mist)
M09	Coolant off
M10	Automatic clamp
M11	Automatic unclamp
M12	Synchronize multiple axes
M13	Spindle clockwise and coolant on
M14	Spindle counterclockwise and coolant on
M15	Rapid motion positive direction
M16	Rapid motion negative direction

(continued)

Table 19-1 *(continued)*

Code	Description
M17–18	Unassigned
M19	Spindle orient and stop
M20–29	Unassigned
M30	End of tape (will rewind tape automatically)
M31	Interlock bypass
M32–39	Unassigned
M40–46	Gear changes if used; otherwise unassigned
M47	Continues program execution from start of program
M48	Cancel M47
M49	Deactivate manual speed or feed override
M50–57	Unassigned
M58	Cancel M59
M59	RPM hold
M60–99	Unassigned
Other Addresses	
A	Rotary motion about the x-axis
B	Rotary motion about the y-axis
C	Rotary motion about the z-axis
D	Angular dimension around a special axis (also used for a third feed function)
E	Angular dimension around a special axis or special feed function
H	Unassigned
I	X-axis arc center point
J	Y-axis arc center point
K	Z-axis arc center point
L	Unassigned
O	Used on some controllers in place of N address of sequence numbers
P	Special rapid traverse code, or third axis parallel to the x-axis
Q	Special rapid traverse code, or third axis parallel to the y-axis
R	Special rapid traverse code, or third axis parallel to the z-axis (also used for radius designation)
U	Secondary axis parallel to x-axis
V	Secondary axis parallel to y-axis
W	Secondary axis parallel to z-axis

Figure 19-10 shows a two-axes program for the lathe that can be used to turn down to a specific size. It can then put threads on the end.

Machining Centers

The concept behind machining centers is having a single machine do as many machining operations as possible. The theory behind this is that, on the average, a part spends more than 90 percent of its manufacturing time waiting to be processed by different machines. Therefore, the longer a part can be processed by a single machine, the greater the reduction in shipping and handling and overall machining time. Moreover, a reduction in shipping and handling also reduces the chances for a part to be lost or damaged.

Machining centers are capable of milling, drilling, boring, reaming, countersinking, counterboring, facing, threading, and tapping. They also have up to six axes of movement and contain an automatic tool changer, a tool magazine that can hold a large number of tools, and a tool-change arm that selects tools and removes them from the spindle. Machining centers also contain an index table that allows access to more than one side of the part automatically.

CAD/CAM

The advances in computer technology have dramatically revolutionized the designing and manufacturing of a part. A *computer-aided design/computer-aided manufacturing* (CAD/CAM) system allows a single person to draw a part three dimensionally, select tools for machining it, create tool paths by picking points on the edge of a part, and download it to a machining center for immediate processing (Figure 19-11).

The post processor for a CAD/CAM system can automatically write the CNC program for the machine tool in a matter of seconds. A new product can literally be taken from design board to the shop floor in a matter of minutes. Because CAD/CAM software can automatically create the program with a post processor, a skilled machinist with computer skills is no longer needed. It should be noted, however, that sometimes the software adds additional moves and cuts that can be omitted by an experienced CNC programmer. Removing unnecessary lines of code could save machining time and money if large quantities of parts are being produced. Therefore, there is still a need for at least one experienced machinist with CNC programming skills for each

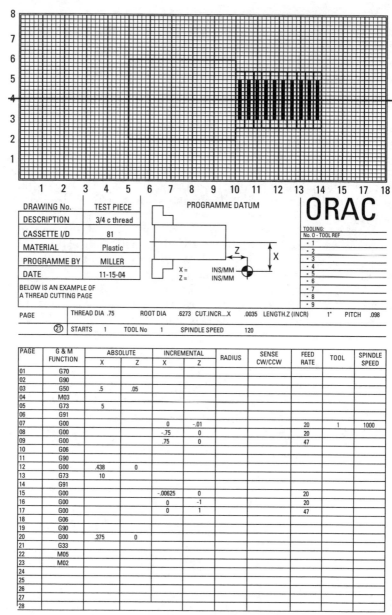

Figure 19-10 Program for cutting a ¾-inch thread.

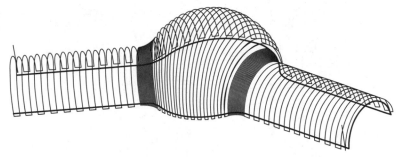

Figure 19-11 NCI file for tool path is written and displayed.

company that produces large numbers of parts using CNC equipment.

Most CAD/CAM software packages have the ability to convert traditional drawing files from familiar drafting software packages such as AutoCAD, MicroStation, and CADKEY. With this capability, designers familiar with a specific software package do not have to waste time learning to draw with their new CAD/CAM system. The designer merely has to learn how to convert files, select post processors, pick points for tool paths, and pick cutting tools. In this way, downtime for training is kept to a minimum and CNC programs can then be produced in minutes instead of days.

Computer-Integrated Manufacturing (CIM)

Computer-integrated manufacturing (CIM) is the term for a manufacturing facility that has all of its functions monitored by a main computer (Figure 19-12). This includes the business functions as well as the machine tools on the shop floor. In this way, a company knows exactly how many parts it is producing, how much material it has in inventory, and which machines are producing to capacity. This allows the company to keep better track of its inventory so that it does not have to spend extra money on unnecessary storage facilities. CIM facilities also have robots and *automatically guided vehicles* (AGV) to handle the loading and unloading of parts. The main purpose of CIM is to transform designs and materials into products for the consumer in the shortest amount of time.

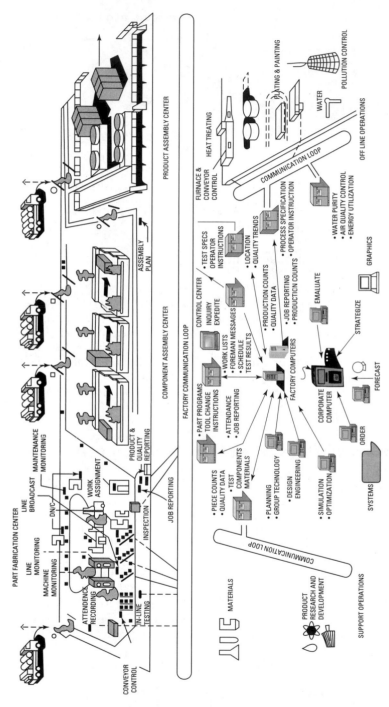

Figure 19-12 CIM. Factory of the future. *(Courtesy of GCA Corporation)*

Summary

Most new machine tools are now computer controlled. Milling machines, drill presses, grinders, turret punches, welding machines, and lathes are some of the more common types of CNC machine tools.

Computer numerical control (CNC) means that a computer controls the movements of a machine by computer programs that are made up of letters, symbols, and numbers. Skilled machinists are no longer needed to run the CNC machines. The machining is done automatically according to the information that is recorded on the program. However, skilled machinists are needed to write the programs to make sure the work is done correctly.

CNC programs can be stored on floppy disks, magnetic tape, or any of the other available media. Programs have the capability of being returned quickly to the machine for reuse. This eliminates the need to replan the job.

CNC machines are more complex and are therefore more expensive to purchase. Costs associated with maintenance are higher as well because of their increased complexity.

Low-cost, full-featured CNC controls specifically designed for mills and machining centers with up to six axes can be retrofitted to new and old manually controlled machines. This unit can be integrated to new or existing machines quickly and easily. It uses a new type of connector to allow controls to be connected without the use of special tools.

Computer control of machine tools became feasible around the mid-1960s with the introduction of the integrated circuit. Computers in the late 1960s were still quite large and expensive compared with today's standards. Because of this, direct numerical control (DNC) became the first type of computer control program. A DNC system is made up of one large computer that can operate many machine tools on a time-shared basis.

In the early 1970s, computers became a lot less expensive and more compact. They were also rugged enough to be subjected to harsh environments that may be encountered on a shop floor. Today, CNC machine tools can be linked to a mainframe computer. This combination (commonly called *distributed numerical control*) is an updated version of direct numerical control. A distributed numerical control system has all the advantages of the earlier direct numerical control system along with the ability to communicate with the CNC machine tools on the shop floor. The two main components of a CNC machine are the machine tool and the machine control unit (MCU). The MCU is the computer that operates the machine tool. There are many manufacturers of these control units, so program codes may vary somewhat between machines.

There are three types of control systems for NC and CNC machines. The least-expensive type is called a point-to-point (PTP) control system. A straight-cut control system is capable of moving the cutting tools in a straight line between points. However, only one axis drive motor can move at one time, so arcs and angles are not possible.

The contouring control system is the most expensive and flexible of the three systems. With the recent advances in electronics, the price difference between these systems has narrowed so much that every CN machine is now manufactured with a contouring control system. A contouring system is capable of doing what the other systems can do while also being able to cut arcs and angles.

A CNC program is written using the Cartesian coordinate system. Lathes are programmed two-dimensionally using only the x- and z-axes, while milling machines are programmed three-dimensionally using the x-, y-, and z-axes. CNC machines can be positioned in two ways with respect to the Cartesian coordinate system. They can be positioned in an absolute format or an incremental format.

Conventional NC machines that are not connected to a computer do not have the memory capabilities to run a canned cycle. Every move of the cutting tools would have to be listed coordinate by coordinate. NC machines also lack the power of a computer's memory to calculate arcs and curves instantaneously. CNC machines have the added feature of graphic simulation. This feature allows the programmer to watch a visual display of how the CNC machine will machine the part according to the program that was just created. Any errors in the program will be displayed so that a programmer can debug a program without wasting material or tool bits.

Because of the added feature of a computer's memory, CNC programming is a lot easier and simpler than conventional NC programming. Most CNC machines can be programmed using a standardized list of codes that were developed by the Electronics Industries Association (EIA).

The concept behind CNC machining centers is having a single machine do as many machining operations as possible. The theory behind this is that, on the average, a part spends more than 90 percent of its manufacturing time waiting to be processed by different machines. If one machine can do the job of several machines, then more time can be spent processing parts versus transporting parts. In the end, more parts are produced and fewer parts are lost or damaged in transit. Machining centers are capable of milling, drilling, boring, reaming, countersinking, counterboring, facing, threading, and tapping. They also have up to six axes of movement and contain an automatic tool changer, a tool magazine that can hold a large number of tools, and a tool-change arm that selects tools and removes them from the spindle.

The advances in computer technology have dramatically revolutionized the designing and manufacturing of a part. A computer-aided design/computer-aided manufacturing (CAD/CAM) system allows a single person to draw a part three-dimensionally, select tools for machining it, create tool paths by picking points on the edge of a part, and download it to a machining center for immediate processing. The post processor for a CAD/CAM system can automatically write the CNC program for the machine tool in a matter of seconds. A new product can literally be taken from design board to the shop floor in a matter of minutes.

Computer-integrated manufacturing (CIM) is the term for a manufacturing facility that has all of its functions monitored by a main computer. This includes the business functions as well as the machine tools on the shop floor. CIM facilities also have robots and automatically guided vehicles (AGV) to handle the loading and unloading of parts. The main purpose of CIM is to transform designs and materials into products for the consumer in the shortest amount of time.

Review Questions

1. What is the meaning of the term *computer numerical control*?
2. Why are skilled machinists no longer needed to run the CNC machines?
3. Where are the skilled machinists needed in the machine shop?
4. What are some advantages of CNC?
5. What are some disadvantages of CNC?
6. When did computer control machines become feasible?
7. What made the computer numerical control of machines possible?
8. What does DNC stand for?
9. What is MCU?
10. What is the Cartesian coordinate system?

11. What is a canned cycle?
12. What is an absolute format?
13. What is an incremental format?
14. What is the advantage of CNC over NC?
15. What is a G code?
16. What is a machining center?
17. What can machining centers do?
18. What is CAD?
19. What is CAD/CAM?
20. What is CIM?

Appendix

Reference Materials

This appendix contains useful reference information, including the following:

- Colors and approximate temperatures for carbon steel
- Nominal dimensions of hex bolts and hex cap screws
- Nominal dimensions of heavy hex bolts and heavy hex cap screws
- Nominal dimensions of heavy hex structural bolts
- Nominal dimensions of hex nuts, hex thick nuts, and hex jam nuts
- Nominal dimensions of square-head bolts
- Nominal dimensions of heavy hex nuts and heavy hex jam nuts
- Nominal dimensions of square nuts and heavy square nuts
- Nominal dimensions of lag screws

Colors and Approximate Temperatures for Carbon Steel

Color	Temperature in Degrees Fahrenheit	Temperature in Degrees Celsius
Black red	990	532
Dark blood red	1050	566
Dark cherry red	1175	635
Medium cherry red	1250	677
Full cherry red	1375	746
Light cherry, scaling	1550	843
Salmon, free scaling	1650	899
Light salmon	1725	941
Yellow	1825	996
Light yellow	1975	1080
White	2220	1216

Nominal Dimensions of Hex Bolts and Hex Cap Screws

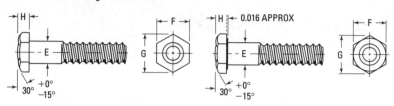

Hex Bolts		Hex Cap Screws	
Nominal Size E	*Width Across Flats F*	*Width Across Corners G*	*Head Height H*
$\frac{1}{4}$	$\frac{7}{16}$	$\frac{1}{2}$	$\frac{11}{64}$
$\frac{5}{16}$	$\frac{1}{2}$	$\frac{9}{16}$	$\frac{7}{32}$
$\frac{3}{8}$	$\frac{9}{16}$	$\frac{21}{32}$	$\frac{1}{4}$
$\frac{7}{16}$	$\frac{5}{8}$	$\frac{47}{64}$	$\frac{19}{64}$
$\frac{1}{2}$	$\frac{3}{4}$	$\frac{55}{64}$	$\frac{11}{32}$
$\frac{5}{8}$	$\frac{15}{16}$	$1\frac{3}{32}$	$\frac{27}{64}$
$\frac{3}{4}$	$1\frac{1}{8}$	$1\frac{19}{64}$	$\frac{1}{2}$
$\frac{7}{8}$	$1\frac{5}{16}$	$1\frac{33}{64}$	$\frac{37}{64}$
1	$1\frac{1}{2}$	$1\frac{47}{64}$	$\frac{43}{64}$
$1\frac{1}{8}$	$1\frac{11}{16}$	$1\frac{61}{64}$	$\frac{3}{4}$
$1\frac{1}{4}$	$1\frac{7}{8}$	$2\frac{11}{64}$	$\frac{27}{32}$
$1\frac{3}{8}$	$2\frac{1}{16}$	$2\frac{3}{8}$	$\frac{29}{32}$
$1\frac{1}{2}$	$2\frac{1}{4}$	$2\frac{19}{32}$	1
$1\frac{3}{4}$	$2\frac{5}{8}$	$3\frac{1}{32}$	$1\frac{5}{32}$
2	3	$3\frac{15}{32}$	$1\frac{11}{32}$

Nominal Dimensions of Heavy Hex Bolts and Heavy Hex Cap Screws

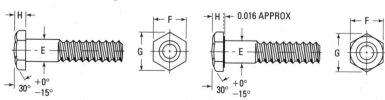

Heavy Hex Bolts		Heavy Hex Cap Screws		
	Width		Height	
Nominal Size	Across Flats **F**	Across Corners **G**	Bolts **H**	Screws **H**
$\frac{1}{2}$	$\frac{7}{8}$	1	$\frac{11}{32}$	$\frac{5}{16}$
$\frac{5}{8}$	$1\frac{1}{16}$	$1\frac{15}{64}$	$\frac{27}{64}$	$\frac{25}{64}$
$\frac{3}{4}$	$1\frac{1}{4}$	$1\frac{7}{16}$	$\frac{1}{2}$	$\frac{15}{32}$
$\frac{7}{8}$	$1\frac{7}{16}$	$1\frac{21}{32}$	$\frac{37}{64}$	$\frac{35}{64}$
1	$1\frac{5}{8}$	$1\frac{7}{8}$	$\frac{43}{64}$	$\frac{39}{64}$
$1\frac{1}{8}$	$1\frac{13}{16}$	$2\frac{3}{32}$	$\frac{3}{4}$	$\frac{11}{16}$
$1\frac{1}{4}$	2	$2\frac{5}{16}$	$\frac{27}{32}$	$\frac{25}{32}$
$1\frac{3}{8}$	$2\frac{3}{16}$	$2\frac{17}{32}$	$\frac{29}{32}$	$\frac{27}{32}$
$1\frac{1}{2}$	$2\frac{3}{8}$	$2\frac{3}{4}$	1	$\frac{15}{16}$
$1\frac{3}{4}$	$2\frac{3}{4}$	$3\frac{11}{64}$	$1\frac{5}{32}$	$1\frac{3}{32}$
2	$3\frac{1}{8}$	$3\frac{39}{64}$	$1\frac{11}{32}$	$1\frac{7}{32}$

Nominal Dimensions of Heavy Hex Structural Bolts

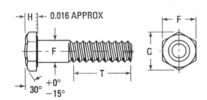

Nominal Size **E**	Width Across Flats **F**	Width Across Corners **G**	Head Height **H**	Thread Length **T**
$\frac{1}{2}$	$\frac{7}{8}$	1	$\frac{5}{16}$	1
$\frac{5}{8}$	$1\frac{1}{16}$	$1\frac{15}{64}$	$\frac{25}{64}$	$1\frac{1}{4}$
$\frac{3}{4}$	$1\frac{1}{4}$	$1\frac{7}{16}$	$\frac{15}{32}$	$1\frac{3}{8}$
$\frac{7}{8}$	$1\frac{7}{16}$	$1\frac{21}{32}$	$\frac{35}{64}$	$1\frac{1}{2}$
1	$1\frac{5}{8}$	$1\frac{7}{8}$	$\frac{39}{64}$	$1\frac{3}{4}$
$1\frac{1}{8}$	$1\frac{13}{16}$	$2\frac{3}{32}$	$\frac{11}{16}$	2
$1\frac{1}{4}$	2	$2\frac{5}{16}$	$\frac{25}{32}$	2
$1\frac{3}{8}$	$2\frac{3}{16}$	$2\frac{17}{32}$	$\frac{27}{32}$	$2\frac{1}{4}$
$1\frac{1}{2}$	$2\frac{3}{8}$	$2\frac{3}{4}$	$\frac{5}{16}$	$2\frac{1}{4}$

Nominal Dimensions of Hex Nuts, Hex Thick Nuts, and Hex Jam Nuts

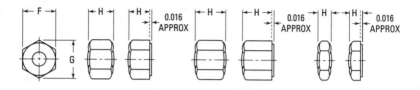

				Thickness	
Nominal Size	Width Across Flats F	Width Across Corners G	Hex Nuts H	Thick Nuts H	Jam Nuts H
$5/16$	$1/2$	$9/16$	$17/64$	$21/64$	$3/16$
$3/8$	$9/16$	$21/32$	$21/64$	$13/32$	$7/32$
$7/16$	$11/16$	$51/64$	$3/8$	$29/64$	$1/4$
$1/2$	$3/4$	$55/64$	$7/16$	$9/16$	$5/16$
$9/16$	$7/8$	1	$31/64$	$39/64$	$5/16$
$5/8$	$15/16$	$1 3/32$	$35/64$	$23/32$	$3/8$
$3/4$	$1 1/8$	$1 19/64$	$41/64$	$13/16$	$27/64$
$7/8$	$1 5/16$	$1 33/64$	$3/4$	$29/32$	$31/64$
1	$1 1/2$	$1 47/64$	$55/64$	1	$35/64$
$1 1/8$	$1 11/16$	$1 61/64$	$31/32$	$1 5/32$	$39/64$
$1 1/4$	$1 7/8$	$2 11/64$	$1 1/16$	$1 1/4$	$23/32$
$1 3/8$	$2 1/16$	$2 3/8$	$1 11/64$	$1 3/8$	$25/32$
$1 1/2$	$2 1/4$	$2 19/32$	$1 9/32$	$1 1/2$	$27/32$

Nominal Dimensions of Square-Head Bolts

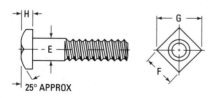

Nominal Size E	Width Across Flats F	Width Across Corners G	Head Height H
$1/4$	$3/8$	$17/32$	$11/64$
$5/16$	$1/2$	$45/64$	$13/64$
$3/8$	$9/16$	$51/64$	$1/4$
$7/16$	$5/8$	$57/64$	$19/64$
$1/2$	$3/4$	$1^{1}/16$	$21/64$
$5/8$	$15/16$	$1^{21}/64$	$27/64$
$3/4$	$1^{1}/8$	$1^{19}/32$	$1/2$
$7/8$	$1^{5}/16$	$1^{55}/64$	$19/32$
1	$1^{1}/2$	$2^{1}/8$	$21/32$
$1^{1}/8$	$1^{11}/16$	$2^{25}/64$	$3/4$
$1^{1}/4$	$1^{7}/8$	$2^{21}/32$	$27/32$
$1^{3}/8$	$2^{1}/16$	$2^{59}/64$	$29/32$
$1^{1}/2$	$2^{1}/4$	$3^{3}/16$	1

Nominal Dimensions of Heavy Hex Nuts and Heavy Hex Jam Nuts

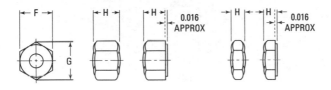

Nominal Size	Width Across Flats F	Width Across Corners G	Thickness Hex Nuts H	Thickness Hex Jam Nuts H
$1/4$	$1/2$	$37/64$	$15/64$	$11/64$
$5/16$	$9/16$	$21/32$	$19/64$	$13/64$
$3/8$	$11/16$	$51/64$	$23/64$	$15/64$
$7/16$	$3/4$	$55/64$	$37/64$	$17/64$
$1/2$	$7/8$	$1^{1}/64$	$31/64$	$19/64$
$9/16$	$15/16$	$1^{5}/64$	$35/64$	$21/64$
$5/8$	$1^{1}/16$	$1^{7}/32$	$39/64$	$23/64$
$3/4$	$1^{1}/4$	$1^{7}/16$	$47/64$	$27/64$
$7/8$	$1^{7}/16$	$1^{21}/32$	$55/64$	$31/64$
1	$1^{5}/8$	$1^{7}/8$	$63/64$	$35/64$
$1^{1}/8$	$1^{13}/16$	$2^{3}/32$	$1^{7}/64$	$39/64$

(continued)

(continued)

Nominal Size	Width Across Flats F	Width Across Corners G	Thickness Hex Nuts H	Hex Jam Nuts H
$1\frac{1}{4}$	2	$2\frac{5}{16}$	$1\frac{7}{32}$	$\frac{23}{32}$
$1\frac{3}{8}$	$2\frac{3}{16}$	$2\frac{17}{32}$	$1\frac{11}{32}$	$\frac{25}{32}$
$1\frac{1}{2}$	$2\frac{3}{8}$	$2\frac{1}{2}$	$1\frac{15}{32}$	$\frac{27}{32}$
$1\frac{5}{8}$	$2\frac{9}{16}$	$2\frac{61}{64}$	$1\frac{19}{32}$	$\frac{29}{32}$
$1\frac{3}{4}$	$2\frac{3}{4}$	$3\frac{11}{64}$	$1\frac{23}{32}$	$\frac{31}{32}$
$1\frac{7}{8}$	$2\frac{15}{16}$	$3\frac{25}{64}$	$1\frac{27}{32}$	$1\frac{1}{32}$
2	$3\frac{1}{8}$	$3\frac{39}{64}$	$1\frac{31}{32}$	$1\frac{3}{32}$

Nominal Dimensions of Square Nuts and Heavy Square Nuts

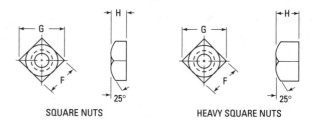

SQUARE NUTS HEAVY SQUARE NUTS

Nominal Size	Square Nuts Width Across Flats Regular F	Heavy F	Width Across Corners Regular G	Heavy G	Thickness Regular H	Heavy H
$\frac{1}{4}$	$\frac{7}{16}$	$\frac{1}{2}$	$\frac{5}{8}$	$\frac{45}{64}$	$\frac{7}{32}$	$\frac{1}{4}$
$\frac{5}{16}$	$\frac{9}{16}$	$\frac{9}{16}$	$\frac{51}{64}$	$\frac{51}{64}$	$\frac{17}{64}$	$\frac{5}{16}$
$\frac{3}{8}$	$\frac{5}{8}$	$\frac{11}{16}$	$\frac{57}{64}$	$\frac{31}{32}$	$\frac{21}{64}$	$\frac{3}{8}$
$\frac{7}{16}$	$\frac{3}{4}$	$\frac{3}{4}$	$1\frac{1}{16}$	$1\frac{1}{16}$	$\frac{3}{8}$	$\frac{7}{16}$
$\frac{1}{2}$	$\frac{13}{16}$	$\frac{7}{8}$	$1\frac{5}{32}$	$1\frac{15}{64}$	$\frac{7}{16}$	$\frac{1}{2}$
$\frac{5}{8}$	1	$1\frac{1}{16}$	$1\frac{27}{64}$	$1\frac{1}{2}$	$\frac{35}{64}$	$\frac{5}{8}$
$\frac{3}{4}$	$1\frac{1}{8}$	$1\frac{1}{4}$	$1\frac{19}{32}$	$1\frac{49}{64}$	$\frac{21}{32}$	$\frac{3}{4}$
$\frac{7}{8}$	$1\frac{5}{16}$	$1\frac{7}{16}$	$1\frac{55}{64}$	$2\frac{1}{32}$	$\frac{49}{64}$	$\frac{7}{8}$
1	$1\frac{1}{2}$	$1\frac{5}{8}$	$2\frac{1}{8}$	$2\frac{19}{64}$	$\frac{7}{8}$	1

(continued)

| Nominal Size | Square Nuts | | Heavy Square Nuts | | | |
| | Width Across Flats | | Width Across Corners | | Thickness | |
	Regular F	Heavy F	Regular G	Heavy G	Regular H	Heavy H
$1\frac{1}{8}$	$1\frac{11}{16}$	$1\frac{13}{16}$	$2\frac{25}{64}$	$2\frac{9}{16}$	1	$1\frac{1}{8}$
$1\frac{1}{4}$	$1\frac{7}{8}$	2	$2\frac{21}{32}$	$2\frac{53}{64}$	$1\frac{3}{32}$	$1\frac{1}{4}$
$1\frac{3}{8}$	$2\frac{1}{16}$	$2\frac{3}{16}$	$2\frac{59}{64}$	$3\frac{3}{32}$	$1\frac{13}{64}$	$1\frac{3}{8}$
$1\frac{1}{2}$	$2\frac{1}{4}$	$2\frac{3}{8}$	$3\frac{3}{16}$	$3\frac{23}{64}$	$1\frac{5}{16}$	$1\frac{1}{2}$

Nominal Dimensions of Lag Screws

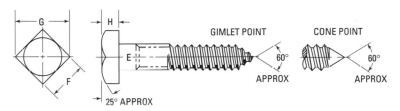

Nominal Size E	Width Across Flats F	Width Across Corners G	Head Height H
10	$\frac{9}{32}$	$\frac{19}{64}$	$\frac{1}{8}$
$\frac{1}{4}$	$\frac{3}{8}$	$\frac{17}{32}$	$\frac{11}{64}$
$\frac{5}{16}$	$\frac{1}{2}$	$\frac{45}{64}$	$\frac{13}{64}$
$\frac{3}{8}$	$\frac{9}{16}$	$\frac{11}{16}$	$\frac{1}{4}$
$\frac{7}{16}$	$\frac{5}{8}$	$\frac{57}{64}$	$\frac{19}{64}$
$\frac{1}{2}$	$\frac{3}{4}$	$1\frac{1}{16}$	$\frac{21}{64}$
$\frac{5}{8}$	$\frac{15}{16}$	$1\frac{21}{64}$	$\frac{27}{64}$
$\frac{3}{4}$	$1\frac{1}{8}$	$1\frac{19}{32}$	$\frac{1}{2}$
$\frac{7}{8}$	$1\frac{5}{16}$	$1\frac{55}{64}$	$\frac{19}{32}$
1	$1\frac{1}{2}$	$2\frac{1}{8}$	$\frac{21}{32}$
$1\frac{1}{8}$	$1\frac{11}{16}$	$2\frac{25}{64}$	$\frac{3}{4}$
$1\frac{1}{4}$	$1\frac{7}{8}$	$2\frac{21}{32}$	$\frac{27}{32}$

Index